STRANGE BEDFELLOWS

STRANGE BEDFELLOWS

How Medical Jurisprudence Has Influenced Medical Ethics and Medical Practice

Ben A. Rich, J.D., Ph.D.

University of California, Davis
Sacramento, California

Kluwer Academic / Plenum Publishers
New York / Boston / Dordrecht / London / Moscow

ISBN 0-306-46665-1

233 Spring Street, New York, New York 10013

http://www.wkap.nl

10 9 8 7 6 5 4 3 2 1

A C.I.P. record for this book is available from the Library of Congress

Printed in the United States of America

To my wife, Kathleen Mills,
and my parents, Ben and Betty Rich

PREFACE

The pervasive influence of law on medical practice and clinical bioethics is often noted with a combination of exasperation and lamentation. Physicians and non-physician bioethicists, generally speaking, consider the willingness of courts, legislatures, and regulatory agencies to insinuate themselves into clinical practice and medical research to be a distinctly negative aspect of contemporary American society. They are quick to point out that their colleagues in other Western developed nations are not similarly afflicted, and that the situation which obtains elsewhere is highly preferable to the legalization and purported over-regulation of medicine that has taken place in the United States during the last fifty years.

In this book I offer a decidedly different perspective. It is, admittedly, not entirely without personal and professional bias. Prior to becoming a full-time academic, teaching bioethics in the setting of an academic medical center, I was, for nearly 20 years, an attorney specializing in health law. Even after earning a doctorate in philosophy, I was frequently considered to be the "resident lawyer" on the bioethics faculty, much more frequently looked to for my insights on the law than my perspective as one who had formally studied moral philosophy and applied ethics. I note this not out of a sense of frustration or disappointment, but as confirmation that even among physicians and non-physician bioethicists, there is widespread recognition that the law does have important contributions to make in assessing the practice of medicine and the conduct of medical research. Prominent bioethics journals recognize the important role that law has played in bioethical decision making by including legal sections in each issue, e.g., *The Cambridge Quarterly of Healthcare Ethics* has "The Caduceus in Court," and the *Hastings Center Report* has "At Law."

The perspective of this book is that, at least from the standpoint of the patient and the research subject, the law has had a decidedly positive influence on medical practice and research. The very concept of informed consent – in clinical practice and human subjects research – was developed and imposed upon those engaged in these medical enterprises by courts. As a result, the very model of the physician-patient relationship was transformed from one of paternalism to one of shared decision making. In other words, through the law the patient was given not merely a voice, but the ultimate authority in determining what may be done by health care professionals in both practice and research settings. Informed consent is the handmaiden of individual autonomy, which became a full-fledged principle of medical ethics only after it had become a principle of medical jurisprudence. Similarly, the concept of prospective autonomy, through which an individual can have an authoritative voice in future health care decisions at a time when he or she has lost decisional capacity, only became a viable concept once legislatures and courts recognized the moral and legal authority of advance directives.

In the early decades after the formulation of informed consent as a right of patients and the subjects of medical research, a curious asymmetry existed. Patients were presumed to have a virtually unqualified right to both give and withhold consent to medical interventions, so long as they were not very important, i.e., not essential to sustaining life or avoiding serious disability. However, in matters of life and death or in the treatment of grave illness, there was a lingering presupposition among medical practitioners that patients had no right to refuse treatment. Such refusals were not considered to be legitimate exercises of individual autonomy, but rather evidence of mental derangement or suicidal behavior. The wave of so-called "right to die" litigation in the United States in the 1970's and 1980's was necessary in order to definitively establish that patients have the right not only to decide about merely elective medical procedures, but those necessary to, quite literally, save life and limb. While there are those who suggest that such a right was implicit in the bioethical principle of respect for patient autonomy, I take the position that not until the courts interceded in case after case, invoking both common law and constitutional principles, did patients actually achieve this final authority over their medical management.

This account of the influence of law on the ethics of medical practice is not, however, unmindful of the shortcomings and adverse sequelae of some legal decisions, statutes, and regulations. There are some common law and constitutional cases in which the sometime admonition of my beloved torts professor in law school, Arno Becht, is particularly apt: "Read it and weep!" But in my judgment, the plight of patients and research subjects would be materially worse if the law had taken a hands off approach.

Toward the end of the book I address the issue of whether, in the new century upon which we have now embarked, the law, as instantiated in court decisions, statutes, and administrative regulations, might be well advised to cede more of a role to clinical bioethics in shaping the ethical parameters of medicine. While I believe there is much to be gained by a more synergistic and cooperative role between law and bioethics, I am quite reluctant to advocate for a diminution in the role of law. As I explain at greater length in the final chapter, bioethics is not well positioned to vindicate such principles as patient rights. Bioethics, as an emerging profession, has assiduously avoided taking on being shackled with the mantle of "moral police" in the clinical setting, and in some instances it has actually turned a blind eye to major ethical shortcomings. Many bioethicists, particularly those without tenured faculty positions, are beholden to institutional administrators and the institutional medical staff for their ability to earn a living. Discretion being the better part of valor, they have, sometimes excessively, sought to accommodate prevailing practices rather than to challenge them on ethical grounds. Furthermore, it remains an undisputed fact of the matter that the law has teeth and clinical bioethics does not. Whereas we can all aspire to change problematic professional attitudes and practices through moral suasion, when time is of the essence or key individuals or groups remain intransigent, nothing short of the force of civil or criminal penalties will suffice. When it comes to those matters about which we as a society will not accept merely episodic or partial compliance with core principles and values, we look to the law to motivate most persuasively.

ACKNOWLEDGMENTS

When one comes, as I have in this instance, as circuitously and incrementally to the production of a book, the list of individuals to whom one is indebted in material ways becomes extensive. Because much of the inspiration for this book comes from my experience as a practitioner of the law, the sources of my motivation to pursue the law as my initially chosen career may be worthy of note. I grew up in central Illinois, having been born in Springfield and raised in Peoria. In high school and college I came to recognize a triumvirate of heroes, all of whom were attorneys and each of whom rose to national prominence in Illinois. Perhaps still more importantly, however, these were individuals who inspired me because they exemplified profound moral courage and an unqualified pursuit of truth and justice regardless of the adversity and animosity which that pursuit invariably engendered in many of their fellow citizens. The first of these, both chronologically and in terms of his national stature, was Abraham Lincoln. While time has conferred upon him the status of perhaps the most revered of American presidents, he was also one of the ablest trial lawyers of his time.

The second, and probably the most obscure, was Robert Green Ingersoll, who lived and practiced law in Peoria from 1857 until he moved to New York in the 1870's. Although he, too, was in the upper echelon of trial advocates in the last half of the Nineteenth Century, Ingersoll's fame (or in the eyes of the religious, infamy) derived from his performance on the lecture circuit, where he was known as the "golden tongued agnostic," and his delivery at the Republican National Convention of 1876 of the "Plumed Knight Speech" nominating James G. Blaine to be the Republican candidate for President.

The third member of the triumvirate, and the one best known as a paradigm of trial advocacy, was Clarence Darrow of Chicago. My admiration for Darrow was germinated as a young adolescent when director Stanley Kramer brought the dramatization of the "Scopes Monkey Trial" to the screen in the 1960 production of *Inherit The Wind*, casting Spencer Tracy as the aging Clarence Darrow, who came to Dayton, Tennessee in the sweltering summer of 1925 to defend James T. Scopes against charges that teaching the theory of evolution in a public school classroom was a crime. Like the other two, Darrow was an individual of unwavering conviction, who achieved greatness in his chosen profession without compromising his principles or tempering his views of what justice required by way of commitment and sacrifice.

In moving from the pantheon of heroes to the figures in my actual experience, I wish to begin with Edward E. Hollowell, Esq., of Raleigh, N.C. for bringing me into the private practice of law and introducing me to the speciality of health law, with its many fascinating issues. I am also most grateful to Gene Nichol, formerly Dean of the University of Colorado School of Law and now Dean of the University of North Carolina School of Law, and to Rex Perschbacher and Kevin Johnson, Dean and Associate Dean respectively at the University of California, Davis School of Law, for affording me the opportunity to teach health law at these fine institutions during the last 12 years. It was in my early years of teaching that I realized that a second career as an academic had begun to beckon.

When I sought to move from the practice of law to full-time university teaching, I was motivated to pursue formal credentials in moral philosophy and applied ethics. I was reassured that such a radical career move, including a return to the role of graduate student, was within the realm of possibility by Professor Frank Marsh, who was then at the University of Colorado Health Sciences Center and now is at the University of Tennessee. Like myself, Frank had been an attorney for many years before pursuing a doctorate in philosophy and embarking upon his second career as an academic bioethicist. My sage mentor and friend in this endeavor was Professor N. Ann Davis, who further encouraged my enrollment in the Ph.D. program of the Department of Philosophy at the University of Colorado-Boulder, and who served as chair of my dissertation committee.

By the time I was nearing the completion of my dissertation, I had the good fortune to have become a colleague and friend of Dr. Mark Yarborough, Director of the Program in Health Care Ethics, Humanities and Law at the University of Colorado Health Sciences Center. As that program was expanded through the support and nurturing of Dr. Vincent Fulginiti, Chancellor of the Health Sciences Center, Mark brought me into the academic fold and tolerated with sublime equanimity my eventual status as the "resident curmudgeon." I also wish to acknowledge the other faculty of that program, Dr. John Armstrong, Dr. Richard Martinez, Dr. Therese Jones, and Dr. Marilyn Coors, for providing a unique admixture of intellectual inquiry, collegiality and friendship that now spans both time and distance.

I have recently completed my first year as Associate Professor in the Bioethics Program at the University of California, Davis. I am particularly indebted to my mentor, friend and colleague Dr. Erich H. Loewy, the Alumni Association Endowed Chair of Bioethics, who encouraged me to come to UC-Davis and to complete this book project. I also wish to thank Dean Joe Silva of the School of Medicine, Dr. Faith Fitzgerald, Dr. Fred Meyers, Dr. Scott

Fishman, and Dr. Roberta Springer Loewy for helping me to thrive in my new academic home.

There are, of course, those to whom expressions of thanks are woefully inadequate because the degree of indebtedness is too great, hence my dedication of this book to them. My wife, Kathleen Mills, has supported me in the pursuit of my academic dream for these many years in all of the ways – tangible and intangible, obvious and subtle – that only a loving and dedicated partner can. And last, but certainly not least, to my parents, Ben and Betty Rich, who created and sustained a nurturing home in which intellectual inquiry was promoted and dogma disdained.

CONTENTS

Chapter 1

INTRODUCTION

Fifty years ago there were no university undergraduate courses in medical ethics, or graduate courses in philosophy with medical ethics as their subject matter. Indeed, there were not even any required courses on the subject in medical school. Medical ethics was narrowly circumscribed by certain texts, beginning with the Hippocratic oath and corpus of writings, and continuing through Percival's *Medical Ethics* and the first American Medical Association (AMA) Code of Medical Ethics published in 1847. The roots of modern medicine are sunk deeply into the soil of pre-Socratic Ionian natural philosophy. Many of the pre-eminent philosophers of that period, including Thales, Anaximander, Demokritos, and Empedocles, were physicians. Medicine is indebted to this form of natural philosophy for its emergence from the domination of myth and religion and its embrace of reason and the search for understanding through a careful examination of cause and effect. With the rise of Hippocrates (460-370 BC), medicine came to be recognized as a discrete body of knowledge and practice rather than merely a branch of philosophy. While the so-called Hippocratic corpus contains 60 medical works, Hippocrates has come to be most commonly identified with the oath that bears his name.[1]

The persistence and primacy of the Hippocratic Oath suggests two contradictory views of the importance of medical ethics in medical practice. Because the Oath is both a code of conduct and a statement of professional values, its enduring significance provides persuasive evidence that medicine has always been viewed as a moral enterprise, one which holds its practitioners to a high standard of professionalism. However, the fact that much of its language and admonitions were extremely outdated even centuries ago should cause us to wonder why a new oath that comports with contemporary thought, speech and practice has been so long in coming. Surely, if the oath was a strong influence on the practice of medicine, the Hippocratic version would have been seen as sorely inadequate. The image of medical school graduates swearing by Apollo, Asclepius, Hygieia, and Panaceia on one day and swiftly moving into their internships in high-tech hospital units on the next at the very least sends a powerful subliminal message that ethics is a quaint, antiquated notion that has no genuine relevance to modern clinical practice.[2]

Until quite recently, the medical literature on the ethics of the profession was neither expansive, nor current, nor read by any but those who had a particular interest in the history of medicine. The prevailing view on any particular aspect of clinical practice was inculcated in medical students, interns and residents not through reading assignments or classroom lectures or discussion,[3] but through the mentoring and modeling (now referred to by some educators as "the hidden curriculum") that takes place on the clinical rotations through hospital wards and the preceptorships in medical offices and clinics that are an integral part of the clinical (third and fourth) years of medical school, internship, and residency training.[4] One major flaw in this form of pedagogy is that it rarely invites critical inquiry into distinctions between what is customarily done and what ought to be done based upon standards, values, and priorities that reflect the larger society and not merely the insular perspective of the practitioners of the profession.[5] Furthermore, while making hospital rounds or moving from one patient examining room to another in a medical office or clinic, there rarely if ever is either the inclination or the opportunity to engage in extended discussions about the ethical principles and conflicting values at stake in the care of a particular patient.

A 1989 article in the journal *Academic Medicine* characterized the subject of medical ethics, and more particularly its inclusion in the medical school curriculum, as "coming of age."[6] By that time, one-third of all U.S. medical schools had at least one required ethics course in the formal curriculum, and by 1994 every medical school had incorporated the teaching of medical ethics in some form into its graduation requirements.[7] A still more recent trend has been away from a stand-alone medical ethics course to the incorporation of medical ethics and humanities components into larger, longitudinal courses that span several years of medical school.[8] A singular advantage of this approach to infusing the medical humanities into the formal medical school curriculum is that the humanities can, by direct association in the longitudinal curriculum with more traditional medical course offerings, be properly portrayed as an essential dimension of sound professional practice. Another significant aspect of the infusion of ethics and the medical humanities into the formal medical school curriculum has been the incursion of non-physician medical ethicists into the teaching of these subjects.

Another quiet revolution in medical education has been the expansion of medical jurisprudence. While the first professorship in medical jurisprudence in the U.S. was established at the College of Physicians and Surgeons in New York in 1814, and another in legal medicine was created at Harvard in 1877, the occupants of these positions tended to be exclusively interested in the field of death investigation. In a 1942 paper, for example, the holder of the Harvard Professorship in legal medicine wrote: "Legal medicine is ordinarily defined as the application of medical knowledge to the needs of justice. Although by definition this would appear to be a broad and scientifically heterogenous field, the practice of legal medicine is concerned chiefly with what might be most adequately described as forensic medicine."[9] Recognition of the need for a much more expansive offering of topics in legal medicine in the medical school curriculum did not come until 1952, when the Committee on Medicolegal Problems of the American Medical Association reported on its survey of medical schools, finding that only 22% had satisfactory instruction in legal medicine.[10] What the committee deemed adequate, and recommended as a required course in the fourth year of medical school, was not a dramatic break with past notions of legal medicine, inasmuch as 11 of the recommended 20 hours in the course were to be devoted to forensic science, while only one each would be devoted to law and the legal system and to public health and regulation.

In 1976, the American College of Legal Medicine developed a recommended medical-legal curriculum for medical schools which included the following topics: consent, malpractice and professional liability, law and the legal system, the physician-patient relationship, confidentiality, emergency treatment, informed consent, and the right to refuse treatment.[11] Since then, the number of medical schools reporting that they offer a course in legal medicine increased from 73 in 1978 to 90 in 1992, although less than one-third required students to take it, as recommended by the 1952 AMA committee.[12]

If we move from the education of physicians and other health care professionals to the law school of 50 years ago, the subject of human values in medicine and its relationship to the law will not be found in the curriculum. One might find a course on forensic medicine or medical jurisprudence, which would in all likelihood focus exclusively on such topics as medical evidence in the trial of cases. Indeed, the evolution of early courses in medical jurisprudence into courses in legal medicine, health law, and bioethics and the law can be traced in the four decades (and five editions) of what was one of the earliest and for a long time the pre-eminent casebook, originally edited by William J. Curran of Harvard. The first edition, published in 1960 under the title of *Law and Medicine*, dealt primarily with the application of medical science to law, particularly in personal injury litigation. The second edition in 1970, coinciding with the proliferation of medical malpractice litigation, incorporated materials in that field as well as the forensic sciences in criminal justice administration. In 1982 the third edition included the emerging field of biomedical ethics and the proliferation of judicial decisions on ethically-charged issues such as abortion, the refusal of life-sustaining treatment, and the determination of death. The fourth edition's expansion of topics reflected the increasing concerns about health care costs and how they might be regulated and to some extent controlled. Seeking to accurately represent its extensive topical coverage, the title ballooned to that of *Health Care Law, Forensic Science, and Public Policy* in the fourth edition in 1990. By that time, the casebook had a number of competitors, one of which was entitled *Bioethics and Law.*

Courses in health law or law and medicine have begun to proliferate throughout law schools, and some institutions have even developed health law institutes and programs that enable interested law students to establish a quasi-major in this burgeoning field of law practice. In its fifth edition in 1998, preceded by the death of William Curran, the title of the book has been shortened to a more manageable *Health Care Law and Ethics*, but its coverage reflects the ever-increasing breadth and complexity of the many ways in which law, medicine, and public policy come together.[13]

The role of law, especially common law and Constitutional law adjudication, in shaping medical ethics and medical practice, has gone without much extensive examination. A clear indication of the significance of that role can readily be discerned by even a cursory review of the 150th anniversary edition of the AMA Code of Medical Ethics, published in 1997, which also contains the current opinions of the Council on Ethical and Judicial Affairs with annotations. The vast majority of annotations are references to judicial opinions and law journal articles, rather than to articles in medical or bioethics journals.

The subtitle of David J. Rothman's previously quoted book, *Strangers at the Bedside,* reveals its subject matter – *A History of How Law and Bioethics Transformed Medical Decision Making.* Rothman is a historian, and his book is a detailed and incisive historical account of the process by which "strangers" to the medical profession, including but not limited to legislators, attorneys, academic philosophers interested in applied ethics, and health care administrators of various stripes, became not only interested but also involved in the policies and practices shaping the delivery of patient care and the decision making attendant to that care.

Still more recently, law professor Roger Dworkin wrote *LIMITS – The Role of the Law in Bioethical Decision Making.*[14] Early in the book Dworkin observes:

> Partly because of the desire for speedy, certain answers; partly because of a popular lack of understanding of the relationship between law and ethics; partly because of ignorance about law and the legal system; and partly because of the belief that law is magic, concern for medical ethics has often become a plea for medical law . . . The law is asked and often tries to resolve questions of medical ethics.
>
> This is not improper. Real questions about real people and about public policy are posed by biomedical development. Real disagreements exist. The law is a primary vehicle for resolving disagreements about public policy and the treatment of real persons. To suggest that the law has no

> role to play in the area of biomedical advance would be both stupid and unrealistic. Yet blind faith in the law's ability to resolve bioethical problems or unthinking acquiescence in the dominant role of law would be equally unsound. Much of the law that exists today is ill advised.[15]

Given the absence of medical ethics and medical law from the training of the majority of physicians in practice today that we have previously noted, it should come as no surprise to us that there is "a popular lack of understanding of the relationship between law and ethics [and] ignorance about law and the legal system [and] belief that law is magic." *LIMITS* surveys at great length and in much detail the developments in the law on certain biomedical issues – abortion, sterilization, alternative reproductive techniques, genetics, death and dying, human subjects research – and moves inexorably from Dworkin's original position to his ultimate conclusion that "the soundest response to a social issue posed by biomedical advance is to begin by assuming that no legal response is necessary."[16] In the process, however, the much needed further inquiry into "the relationship between law and ethics" is not undertaken, nor are we told precisely on what grounds or in what manner courts are supposed to refuse to decide cases raising "real questions about real people [with] real disagreements" when all of the requisite jurisdictional prerequisites for adjudication have been met.

In certain respects, therefore, the thesis that will be proposed and evaluated in this book is the opposite of Dworkin's, i.e., that the law has generally played a positive role in the development of our bioethical thinking about certain critical issues in patient care, and that it is a role which the very nature of the American legal system supports, if not requires. The role and influence of law in health care generally and medical practice in particular, especially during the last fifty years, has been substantial. Therefore, it will be necessary to limit our analysis to a few select areas. The criterion for doing so will be ethics. Those topics in medical jurisprudence that involve the most contentious ethical issues are informed consent, advance directives, contraception and abortion, end-of-life care, and physician-assisted suicide. In making these the focus of the book, it is important to recognize that our attention will be focused on ethical and legal rights and duties *in* health care, not ethical and legal rights and duties *to* health care. While the United States may be in the vanguard of the former, it is most assuredly bringing up the rear with regard to the latter.

In the process of explicating the thesis that law has had a generally salutary influence on the ethics of patient care, we will also note the instances in which courts and/or legislatures have fallen short. In some instances, higher courts or subsequent legislative action have dealt with the problems, in others they have not. The relationship between law and ethics will be a continuing concern and consideration. At the outset, therefore, it is important that we achieve clarity on a number of terms, concepts and principles that are integral to medical jurisprudence and medical ethics.

ENDNOTES AND REFERENCES

1. Despite the familiarity of the name 'Hippocrates' and the traditional association of the Hippocratic Oath with the core values of the profession of medicine, it is extremely difficult to separate historical fact from myth and legend. It has been observed that there is so little known about him that his name is one "lacking any accessible historical reality." John Walton, Jeremiah A. Barondess, Stephen Lock, eds., The Oxford Medical Companion (Oxford: Oxford University Press, 1994), p. 370. The authorship of the oath is also the subject of significant historical skepticism, with one scholar contending that the oath's admonitions against abortion and euthanasia are conclusive evidence that it could not have been either written or subscribed to by Hippocrates or the majority of physicians practicing in the Greek world of that time, since both practices were common and well-accepted. Ludwig Edelstein, *The Hippocratic Oath: Text, Translation, and Interpretation* (Baltimore: Johns Hopkins University Press, 1943), p. 10. In tracing the history of abortion, the Opinion of the U.S. Supreme Court in *Roe v. Wade* pays particular attention to Edelstein's view. 410 U.S. 113, 130-132 (1973).
2. While some share this concern and have called for a new oath that might aspire to the same level of general acceptance within the profession, see Roger J. Bulger, "The Search for a New Ideal," in Roger J. Bulger, ed., *In Search of the Modern Hippocrates* (Iowa City: University of Iowa Press, 1987), pp. 9-10, others continue to defend the Oath in its entirety while criticizing modern codes of ethics such as that of the American Medical Association, see Leon R. Kass, *Toward a More Natural Science* (New York: Free Press, 1985), pp. 224-246.
3. In 1972, for example, only 4% of U.S. medical schools taught medical ethics in a formal, required course. Robert M. Veatch, S. Sollitto, "Medical Ethics Teaching Report of a National Medical School Survey," *Journal of the American Medical Association* (1976); 235: 1030-33.
4. Frank W. Hafferty and R. Franks, "The Hidden Curriculum, Ethics Teaching, and the Structure of Medical Education," *Academic Medicine* (1994); 69: 861-871.
5. As David Rothman has pointed out, the "powerful tradition of ethical discourse . . . was at once high-minded, generous, and even heroic, yet remarkably insular and self-serving, too. Physicians almost exclusively defined the problems and arrived at the resolutions, giving the deliberations a self-contained quality." David J. Rothman, *Strangers at the Bedside* (New York: Basic Books, 1991), pp. 101-102.
6. Steve Miles, et al., "Medical Ethics Education: Coming of Age," *Academic Medicine* (1989); 64: 705-714.
7. Ellen Fox, Robert M. Arnold, Baruch Brody, "Medical Ethics Education: Past, Present, and Future," *Academic Medicine* (1995), 70: 761-769.
8. For a discussion of some of the most comprehensive of such courses see Raymond H. Curry and Gregory Makoul, "The Evolution of Courses in Professional Skills and Perspectives for Medical Students," *Academic Medicine* (1998); 73: 10-13.
9. Alan R. Moritz, "The Need of Forensic Pathology for Academic Sponsorship," *Archives of Pathology* (1942); 382-386.

10. Leslie J. Regan, et al. "Report of the Committee on Medicolegal Problems of the American Medical Association: A Suggested Course in Legal Medicine for Medical Schools," *Journal of the American Medical Association* (1952); 150: 716.
11. Task Force on Medical Legal Curriculum, "A Recommended Medical-Legal Curriculum," *Journal of Legal Medicine* (1977); 5: 8EE-8FF.
12. Peter C. Williams and William Winslade, "Educating Medical Students about Law and the Legal System," *Academic Medicine* (1995); 70: 777-786.
13. William J. Curran, *Health Law and Ethics* (N.Y.: Aspen Law and Business, 1998).
14. Roger B. Dworkin, *LIMITS – The Role of Law in Bioethical Decision Making* (Bloomington, IN: Indiana University Press, 1996).
15. Dworkin, note 14, p. 2.
16. Dworkin, note 14, p. 169.

Chapter 2

MEDICAL ETHICS AND MEDICAL JURISPRUDENCE: THE CONCEPTUAL LANDSCAPE

1. ETHICS

There is currently much debate in the bioethics community about the best way to "do" ethics in the health care setting. In part this debate has been fomented, or at least broadened and intensified, by the socio-political movements of feminism and multiculturalism. These movements challenge ethical norms and standards of professional conduct that were formulated in an earlier era when women and racial minorities had essentially no role in Western medicine or, for that matter, Western social and political philosophy. In order to understand and appreciate these recently developed responses to what formerly was the prevailing view, we must first become familiar with the key elements of that traditional view.

1.1 Professional Competence: Requisite Knowledge and Skill

The *sine qua non* of the "good" physician is technical competence. The essence of any professional, including a physician, is that they possess the fundamental knowledge and skills that are deemed essential by those who set the standards of the profession. To hold oneself out to the public as possessing knowledge and skills which one in fact does not is to "profess" falsely and hence to perpetrate a fraud upon the community and the larger society. So basic is the requirement of technical competence that it is often omitted from discussions of professional ethics, since at least in theory one cannot enter any profession today without having demonstrated in a variety of ways mastery of the requisite knowledge and skills. The operative phrase is "in theory," since in the course of this book we will encounter many examples in which it is quite clear that a particular physician or other health care professional lacks some basic prerequisite which they were presumed to possess. With the vast and rapid expansion of knowledge, particularly in the biological and medical sciences, it becomes increasingly important that health professionals recognize their limitations as a guiding ethical principle and work within them while at the same time seeking to overcome them.

1.2 Beneficence and Nonmaleficence

Possessing and applying a level of knowledge and skill deemed to be minimally sufficient by those who set the standards for professional competence is the surest means of complying with one of the most basic principles of medical ethics – nonmaleficence, or the avoidance of harm. Whether nonmaleficence should be recognized as a separate principle – the opposite side of the coin from beneficence (to act for the benefit of another) – is also a matter of continuing dispute in ethics. As Beauchamp and Childress point out in their treatise on biomedical ethics, many people think of nonmaleficence as the first principle of medical ethics because they associate it with an aphorism attributed to Hippocrates: *Primum non nocere*: "Above all [or first] do no harm."[1] While that precise language cannot be directly linked to the Hippocratic corpus of writings, the spirit of both beneficence and nonmaleficence are expressed in the portion of the Hippocratic Oath which reads: "I will use treatment to help the sick according to my ability and judgment, but I will never use it to injure or wrong them."

The *raison d'etre* of any profession is some form of public service, with the possible exception of the clergy, who typically view their highest calling as service to God and in bringing those in their congregation to a closer relationship with God. In the case of medicine and the health professions, public service primarily takes the form of the care of patients.

As traditional medical ethics has evolved into bioethics during the last half of the Twentieth Century, one of the critical issues that has come to the forefront is who should be the final arbiter of what constitutes harm and benefit to a particular patient. When the power and sophistication of medical science were very limited, and medical technology nonexistent outside of the contents of the physician's proverbial "black bag," physicians were presumed to know the boundaries of harm and benefit, and this presumption was noncontroversial. Today, we find numerous instances in which patients and health care professionals are at odds over what the honoring of the principles of nonmaleficence and beneficence actually means in the provision of care and treatment.

1.3 Paternalism

The increasing frequency and stridency of these disputes has resulted in resort to the courts, by both health care professionals and patients, at least in the United States. The pursuit of a legal resolution to such disputes has also forced a much more thoughtful and deliberate consideration of the principle of respect for patient autonomy. The idea that patients are (except in certain circumstances having to do with mental incapacity) autonomous agents who should be the ultimate arbiters of the care they receive is hardly reflected in early formulations of medical ethics. The explanation is that for many centuries the prevailing model of the physician-patient relationship, as will be considered in depth in Chapter 4, was that of paternalism. Patients were customarily treated by physicians as though they were children, unable to discern what was in their best interests, and encouraged to conduct themselves accordingly. The proper role for the physician was to tell the patient what to do, or what not to do, but not why that was necessary. The proper role of the patient was to comply, without questioning or second-guessing, with the physician's instructions. It is not happenstance that the common term for the physician's instructions is borrowed from the military, i.e., "order." In the Hippocratic era, for example, the physician is admonished to perform his clinical work:

> . . . calmly and adroitly, concealing most things from the patient while you are attending to him. Give necessary orders with cheerfulness and serenity, turning his attention away from what is being done to him; sometimes reprove sharply and emphatically, and sometimes comfort with solicitude and attention, revealing nothing of the patient's future or present condition.[2]

The Hippocratic view prevailed in Western medicine well into the Twentieth Century, with the various editions of the AMA Code of Medical Ethics containing no admonitions with regard to providing information to patients or engaging with them as mature adults about recommended treatment and its risks, benefits, and alternatives. Indeed, the earliest versions of the AMA Code were not original, but were drawn nearly verbatim from Percival's *Medical Ethics*.[3] Among its provisions we find the following:

> Art. I, § 2: Reasonable indulgence should be granted to the mental imbecilities and caprices of the sick.
>
> Art. I, § 4: It is therefore a sacred duty . . . to avoid all things which have a tendency to discourage the patient and to depress his spirit.

> Art. II, § 6: The obedience of a patient to the prescriptions of his physician should be prompt and implicit. He should never permit his own opinions, as to their fitness, to influence his attentions to them.[4]

1.4 Autonomy

The concept of autonomy, despite its extended absence in the traditional discourse of medical ethics, is basic to Western moral philosophy. The etymological roots of the word autonomy can be traced to the Greek *autos* ("self") and *nomos* ("rule"). These roots suggest the essential elements of the autonomous individual. The first is a concept of oneself as a unique person capable of formulating plans, projects, and goals which reflect one's values and priorities. The second is the liberty or freedom to act in a manner that is consistent with those values and objectives. The presence of both elements is generally considered to be prerequisites for moral agency, which entails responsibility for one's actions. One who does not have a developed concept of self, either because of immaturity or mental disorder, cannot formulate the intent to engage in an action that would justify holding them accountable for its consequences. Similarly, one who's actions are constrained or coerced in such a way that they cannot act consistent with their genuine intent cannot be said to have acted, or in the case of an omission, not to have acted freely.

With regard to medical ethics, patient autonomy can be constrained in a number of ways. One of the more obvious would be placing a patient in physical restraints so that he or she may not move about at will. Patients may be ethically restrained in this fashion only when there are good reasons to believe that they pose an imminent danger to themselves or others as a result of some mental disease or defect. The mere fact that a patient refuses to comply with a physician's orders would not be sufficient so long as the patient has decisional capacity, i.e., the ability to understand and appreciate the risks of refusing treatment. A more subtle restraint takes the form of incomplete or inadequate information. A patient who agrees to undergo a medical procedure without adequate information about the risks posed by the treatment and the existence of reasonable alternatives to it is not, strictly speaking, acting autonomously. While the restraints (more accurately constraints) are epistemological rather than physical, they are nonetheless real and limiting. In Chapter 4 we will consider at much greater length the relationship between information disclosure and autonomy in health care.

1.5 Justice

The last of the four ethical principles generally considered to be basic to medical practice is justice. It is also the most subtle and elusive. In part this may be because of the historic primacy of the dyadic relationship between physician and patient. So long as the physician acts in a consistently beneficent manner toward each patient, justice does not seem to be implicated. It is when we expand our focus to the larger society, and to problems and disparities in the access of some individuals or groups to health care, that the role of justice becomes clear and significant. But the dichotomy is not as rigid and simplistic as that. Consider the emergency room of a busy urban hospital. The triage system is based upon a particular formulation of the material principle of distributive justice – to each person according to need. The more gravely ill, usually as determined by a triage nurse, are seen first. A similar approach has been urged for the allocation of the extremely scarce organs, e.g., hearts and livers, for transplantation – patients with the gravest prognosis should be highest on the waiting list of recipients. An alternative approach has been that priority on the waiting list should go to those with the greatest likelihood of survival with a transplant. The most gravely ill prior to transplantation are generally not those with the most favorable prognosis post- transplantation. This is the shape the debate has taken at the present time in the United States, with the Department of Health and Human Services (HHS) issuing a proposed final rule governing the operation of the Organ Procurement Transplant Network (OPTN) that would require that organ allocation among patients for whom the organ is a "match" be determined solely by medical emergency, with waiting time being the factor to distinguish among patients in the same category.[5] The United Network for Organ Sharing (UNOS), a private nonprofit organization with which HHS has contracted for the implementation of the OPTN, defends the alternative approach, and argues that the proposed final rule would require most patients to wait until the severity of their illness virtually precluded that a transplant, if and when ultimately obtained, would be successful.

The United States' initial experience with establishing criteria for allocation of a scarce medical resource, kidney dialysis, was an open and notorious failure from an ethical perspective. The material principle of distributive justice applied in that instance – to each person according to merit – was adopted and implemented by a community committee that came to be known caustically as the "God squad" because, as reporter Shana Alexander indicated in the title of her 1962 *Life* Magazine article on the work of the Admissions and Policies Committee of the Seattle Artificial Kidney Center at Swedish Hospital, "They Decide Who Lives, Who Dies." After patients were screened out by physicians on clinical grounds, which included age (no one over 45, no children) and mental stability, the lay committee was left to develop its own criteria for further reducing the number to the level that could be treated with the new technology. The committee, it appears, sought to determine which patients were making, and would continue to make upon receipt of this life-saving technology, the most significant contribution to society.

The over-arching social justice issue in health care is whether or not there is a societal obligation to provide every person who is a member of that society with some minimally sufficient level of care. The vast majority of Western industrialized nations have long ago answered that question in the affirmative, and have implemented some system for providing universal coverage. Indeed, only the United States has steadfastly refused to recognize such a societal obligation.[6]

1.6 Principalism in Bioethics

Primary reliance upon such core bioethical principles as we have just considered, often characterized as "principalism," has been the subject of much recent criticism. It is fair to say that one of the reasons principalism has been singled out for multiple and extensive critiques is its status as the most prominent approach to bioethical issues. Thus, it is not so much the soundness of principalism, but its hegemony in bioethical discourse, that is challenged by those who advocate alternative approaches to bioethics. One obvious problem with principalism is that in the particulars of any given case two principles may be in conflict. In the case of a competent patient who suffers from a life-threatening illness but who declines the medically indicated therapeutic measures, the principle of respect for patient autonomy appears to conflict with the principle of beneficence. To honor the patient's refusal of treatment is to fail to benefit the patient by saving her life, the most fundamental of beneficent acts. To impose treatment against the patient's will or to coerce the patient into consenting to the necessary treatment is to fail completely to respect the dignity and the autonomy of the patient. Resolution of such a dilemma appears to require privileging one of the principles over the other.

One possible response of principalism to this critique is that respect for autonomy and beneficence only appear to conflict when an unjustifiably narrow, medicalized, and hence paternalistic view of "benefiting" the patient is taken. From a more holistic and patient-oriented perspective, respecting the patient's wishes may benefit her in a deeper and hence ultimately more important way. At various points in this book, and particularly in Chapter 8, we will consider the gradual (often grudging) recognition on the part of the health professions that patient benefit must sometimes be assessed and understood in terms other than saving life or maximizing longevity.

Another criticism of principalism is that even when only one principle is relevant to the case under consideration, the generic formulation does not provide sufficient guidance as to how it should be applied in order to arrive at a reasonable resolution of the ethical issue. Something more than the brute acknowledgment of the validity of the principle is required. Beauchamp and Childress, the most notable spokespersons for principalism, acknowledge this. As their treatise has evolved through four editions over 15 years, they have developed a framework for ethical decision making in health care which positions the principles mid-way between an over-arching ethical theory and rules developed from the principles, each of which is essential to the process of ultimately arriving at a judgment in a particular case. Thus a broader ethical theory, e.g., utilitarianism or consequentialism, informs the goals sought in the application of a principle, and rules provide a necessary level of specificity for arriving at a judgment which a bare statement of principle lacks. Beauchamp and Childress go even further and divide rules into three categories: 1) substantive rules, e.g., truthtelling, confidentiality, privacy, fidelity; 2) authority rules, e.g., surrogate, professional, distributional; and 3) procedural rules, e.g., for determining eligibility for scarce medical resources or for pursuing a grievance to a higher authority. Beauchamp and Childress also note that other relevant considerations in bioethical analysis and decision making, such as the rights of persons, the character and virtues of agents, and the role of the moral emotions, should be an aspect of a comprehensive moral theory, and will at times take on a greater significance than principles and rules.[7] A detailed account of the predominant ethical theories is beyond the scope not only of this chapter, but of the book as a whole. Where appropriate, readers will be directed to authoritative sources of information and analysis for each of the theories mentioned. Both in bioethics and medical jurisprudence, there is no single ethical theory that holds sway. Moreover, as Beauchamp and Childress point out: "distinctions between types of theory are not as significant for *practical* ethics as has sometimes been proclaimed. It is a mistake to suppose that a series of continental divides separates moral theorists into distinct and hostile groups who reach different practical conclusions and fail to converge on principles."[8] With this caveat, it will serve our purposes to concisely canvass the more prominent ethical theories mentioned in contemporary bioethical discourse.

1.7 Utilitarianism

The ethical theory that has come to be most closely associated with the Nineteenth Century British philosopher John Stuart Mill, also known as consequentialist or consequence-based theory, is focused upon ends. Such a theory maintains that the moral worth of an action is to be ascertained by the extent to which it yields the best result. The criterion for evaluating a result is characteristically described as "the greatest happiness for the greatest number." More particularly, the increase of pleasure and the reduction of pain are most likely to be conducive to the utilitarian's goal. Mill warns, however, that in the case of persons, happiness must be understood in a deeper and more philosophical sense than mere physical pleasure. Happiness of this type takes on the character of Aristotelian *eudaimonia*, or human flourishing. Such a philosophical view of happiness is also consistent with the use of that term in the foundational documents of the United States, i.e., the Declaration of Independence and the Constitution. Thus we begin with the Declaration's delineation of certain "self-evident truths," among them that "all men are created equal [and] endowed by their Creator with certain unalienable Rights, . . . among these are Life, Liberty and the pursuit of Happiness – That to secure these rights, Governments are instituted among Men . . . ." These unalienable rights are essential elements of the "Blessings of Liberty" which, according to the Preamble to the U.S. Constitution, the latter document was drafted and ratified in order to make secure.

Louis Brandeis, one of the greatest justices ever to sit on the Supreme Court of the United States, elaborated upon these spare texts in the following memorable words:

> The makers of our Constitution undertook to secure conditions favorable to the pursuit of happiness. They recognized the significance of man's spiritual nature, of his feelings and of his intellect. They knew that only a part of the pain, pleasure, and satisfaction of life can be found in material things. They sought to protect Americans in their beliefs, their thoughts, their emotions and their sensations. They conferred, *as against the government, the right to be let alone* – the most comprehensive of rights and the right most valued by civilized men.[9]

The political ideal of life, liberty and the pursuit of happiness as the natural rights of persons are antecedent to and hence not created or conferred upon citizens by any government, was grounded upon the views of natural law widely held by those who signed the Declaration and the Constitution. The single most important philosopher whose work inspired these individuals and thereby informed both the letter and the spirit of these documents was John Locke. This Lockean conception of liberty is emphasized now and discussed in more detail later in this chapter for it will play an important role in understanding both the Constitutional and the Common Law jurisprudence of autonomy and privacy in health care decision making in subsequent chapters.

Utilitarians posit utility as the ultimate standard for moral judgment. However, they disagree upon whether the focus should be upon the consequences of a particular act ("act utilitarians") or upon the more general rule ("rule utilitarians"). Take truthtelling as an example. Rule utilitarians will maintain that if the principle of truthtelling in health care is conducive to the general good of patients, then exceptions may not properly be made in those rare instances in which it will not be likely to yield the best outcome. Act utilitarians, however, will justify making exceptions to the rule in those cases in which utility is better served by something less than full disclosure. We shall explore the two approaches further in Chapter 4 when the exceptions to the requirement of obtaining a fully informed consent are discussed.[10]

1.8 Deontology

The other most prominent and influential ethical theory is deontology (formalist or obligation-based). This theory draws heavily upon the ethical writings of Immanuel Kant, and is often characterized as the opposite of utilitarianism.[11] For Kant, reason and duty circumscribe the moral realm. The only thing of moral value is a good will, and a good will is one which knows what duty requires (that which reason commands) and acts for the sake of that duty. Thus, a Kantian ethic does not concern itself with the consequences of an action, nor would it accept as moral an act which happens to conform with duty, but which was performed out of some other motive, such as the avoidance of punishment or the pursuit of public favor or privilege. In Kant's own words:

> A good will is good not because of what it performs or effects, not by its aptness for the attainment of some proposed end, but simply by virtue of the volition, that is, it is good in itself, and considered by itself is to be esteemed much higher than all that can be brought about by it in favor of any inclination, nay, even of the sum-total of all inclinations.[12]

If a good will is one which acts according to duty solely for the sake of duty, then the next logical question is how one is to discern duty, particularly in morally ambiguous circumstances which seem to have proliferated in health care. Kant believed that any rational being had the capacity to discern what the law of reason required in any particular situation, such that a rational agent possessed of a good will can at all times know what duty requires and act in conformance to it. As a product of this analysis, Kant derives the concept of the "categorical imperative," which maintains that one must act only upon a maxim which one can will to become a universal law. A maxim which will not meet the test of universality cannot be a source of duty and hence genuine moral action.

There are, readers will have recognized by now, significant problems with any deontological ethical theory as rigid as Kant's formulation. First, it leaves too much unspecified and thus open to quite different "rational" perspectives. Second, it provides too little guidance in those situations presenting genuine moral dilemmas, i.e., where what seem to be equally compelling ethical obligations are in fact in conflict, such that meeting one ethical obligation necessarily entails the violation of another. Kant might respond that the failure to ascertain which obligation is in fact one's real duty (because it is what the law of reason requires) simply confirms the inadequacies of the agent's faculties of moral discernment. Nevertheless, such situations do seem to arise and confound even the most morally scrupulous. Therefore, a strict Kantian ethics may simply demand too much from most of us, and fail the feasibility test.

1.9 Virtue Theory

Virtue-based ethics begins with the premise that knowing and pursuing the good is not simply a matter of subscribing to the right moral theory or set of principles; it involves in an essential way the character of the individual. If we trace virtue-based theory back to ancient Greece, we find that Plato considered vice (the opposite of virtue) to be a product of ignorance, inasmuch as the virtues were considered to be discoverable by proper exercise of one's rational faculties. Aristotle, on the other hand, considered virtue to be a disposition to act in the proper way. Consequently, Aristotle made an important distinction between an action and the motive for the action. On his account, it is possible for an action to be "right" in the sense that it conforms to the prevailing norms of a society, without being virtuous. To be virtuous, an action must be the product of a virtuous state of mind. In that limited sense, at least, Kantian deontology is consistent with Aristotle in its insistence upon the overriding importance of the subjective state of the actor in making judgments about virtuous conduct. If a person's action is entirely situational, unrelated to any underlying trait of character or core belief in a law-like moral code, there is no reasonable expectation that such a person's future actions will be consistent in other situations that would reasonably require consistent action.

Among contemporary discussions of virtue outside of the context of bioemedical ethics, one of the most influential has been that of Alasdair MacIntyre.[13] MacIntyre defines a virtue as "*an acquired quality the possession and exercise of which tends to enable us to achieve those goods which are internal to practices and the lack of which effectively prevents us from achieving any such goods* (italics in original)."[14] MacIntyre's definition of virtue is particularly suitable to the consideration of the virtues in health care because of its focus on practices, by which he means not merely technical skills, but a broader range of *praxis* and of interrelationships among those who engage in those practices in pursuit of goods that are recognized by a consensus. Moreover, MacIntyre recognizes that the practices in which certain virtues are instantiated take place within and must be supported by institutions. In this regard, he notes that "the ability of a practice to retain its integrity will depend on the way in which the virtues can be and are exercised in sustaining the institutional forms which are the social bearers of the practice."[15] Medicine would seem to offer a paradigmatic case of MacIntyre's conception of the virtues and their interrelationship with practices and the institutions in which those practices take place. The hospital, and more recently the managed care organization (MCO), have a profound influence not only upon the fostering of the virtues in medicine, but also in the process by which something is recognized to be a virtue.[16]

One of the arguments in support of giving the virtues a prominent place in health care ethics is that medicine, and more broadly all of the health care professions, is most essentially a moral enterprise. The end of medicine, from its inception, has been the promotion of the patient's good. It is important to recognize that many of the proponents of a major role for the virtues in biomedical ethics do not suggest that they should be treated as sufficient unto themselves, such that they would obviate the need for the consideration of other ethical theories or principles. What is needed, they insist, is a synthesis of the concern with virtues and the kind of principle-based ethics that Beauchamp and Childress have embraced in the more recent editions of their treatise.[17]

Such an approach tends to weaken one of the persistent critiques of virtue theory generally, which is that while consideration of virtue might be a necessary element of making ethical judgments or of insuring consistently ethical practices, it cannot reasonably be offered as a sufficient one. Virtue theorists, especially in bioethics, tend to agree with this argument, but they maintain that the same can be said about any of the other approaches. Certainly, without some other theoretical foundation, and principles and rules to provide it with depth and complexity, we would lack a noncontroversial basis upon which to discuss coherently what the virtues in health care actually are or should be, or more importantly, in what way they can serve as guides to action in highly ambiguous situations. Among the primary candidates for the virtues of the health care professional are fidelity, compassion, prudence, and integrity.[18] While probably no one would seriously dispute their place in a pantheon of physician virtues, standing alone they provide little direction to practitioners.

1.10 Narrative Ethics

MacIntyre's account of virtue leads quite naturally into a discussion of the role of narrative in ethical judgment and decision making, and enables us to consider under this heading a number of different contemporary approaches to ethics, including community based ethical theory and the ethics of care. The critical nexus among these approaches (a more apt term than theory because of their emphasis on perspective) is that individuals understand themselves and are (or should be) understood by others as beings with a single or unitary life history that is embedded in relationships with others and within a larger community. Thus persons, and more importantly for our purposes, persons who happen to be patients, attempt to understand themselves and their lives by an interlocking set of narratives. MacIntyre argues that "I am never able to seek for the good or exercise the virtues *qua* individual . . . we all approach our own circumstances as bearers of a particular social identity."[19]

Narrative ethics, along with feminist and care ethics, which we will consider next, can in one sense be viewed as reactions to or rejections of certain philosophical positions that predominated in the modern (beginning with the Enlightenment) era. Narrative ethics challenges the position of many prominent analytic philosophers that personal identity and a full understanding of the unity of the life of a person can be understood as nothing more than the persistence of certain psychological connections over time. MacIntyre insists that "all attempts to elucidate the notion of personal identity independently of and in isolation from the notions of narrative, intelligibility and accountability are bound to fail."[20] Such issues of personal identity – what it consists in and under what conditions it is lost – will be of critical importance in the cases discussed in Chapters 5 and 8.

1.11 Feminist and Care Ethics

The feminist critique of social and political institutions has also been brought to bear upon moral philosophy, in the form of feminist ethics, and bioethics, in the form of feminist bioethics. While there is tremendous diversity of perspective and opinion within feminist philosophy, a consensus does seem to have formed which moves beyond feminism's foundational convictions that women generally fare worse than men in most social contexts and that the patent injustice of this state of affairs is an injustice that must not be tolerated. Thus it has been suggested that there are certain core propositions or tenets (feminism eschews traditional Western philosophy's obsession with principles) of feminist ethics. However, these tenets are basically critiques of some of the salient features of Western moral philosophy, e.g., a predilection for abstract theory, a disdain for emotion and empathy in moral reasoning, an insistence that choice is central to morality. However, there is no reason to discount the significance of feminism's foundational concerns. After all, one of the icons of traditional Western philosophy, the philosopher whose ethics has become synonymous with deontology, Immanual Kant, claimed that women lack moral reason.[21]

One dimension of feminist ethics has argued for an "ethics of care." Such an ethic is based upon seminal work by psychologist Carol Gilligan suggesting that the moral development of women is actually different from that of men.[22] The "ethics of care," which for many feminist philosophers should not be viewed as something separate and distinct from feminist ethics, denotes a characteristically feminine approach to morality and ethical reflection that naturally takes into account the needs of others and the emotional bonds between and among persons.[23] Given the prevailing (curative) model of medicine, there is a potentially problematic aspect to the adoption of the term "ethics of care." Because of modern medicine's focus on curing over caring, advocacy of a "care ethic" may, misleadingly, suggest that it denotes nursing ethics. Indeed, some of the literature in feminist bioethics actually facilitates such confusion.[24] The confusion is exacerbated by the suggestion, implicit in the notion of a unique, gender-based way of moral reasoning, that only in health professions where women predominate, such as nursing, can the care ethic reasonably be expected to obtain.

Feminist ethics finds fertile ground for its core critique in medicine and bioethics. Medicine has been and, despite a recent influx of women into medical school classes, in many ways continues to be very much a male-dominated profession. Moreover, modern medicine's obsession with the curative, high technology model of patient care denies the significance of much that feminism emphasizes, such as the contextual, the subjective, emotion and empathy. Feminist bioethics asserts that the field of bioethics itself has been co-opted by medicine, thereby engendering a disturbing degree of moral myopia or indifference to such ethically problematic situations as the millions of people without access to health care, and the relegation of women to subservient roles in the health care professions, with nursing as a prime example.

Many of the major issues in bioethics, including a number that we will consider at length in subsequent chapters of this book, present unique problems for feminism. Most prominent among these is abortion. For example, although the advocates for reproductive liberty include a larger percentage of women and tend to couch their rhetoric in terms of a woman's right to choose, feminist ethics, as we have noted, is uncomfortable with traditional moral philosophy's obsession with individual autonomy. As we shall explore further in Chapter 7, feminist bioethics has also de-emphasized the significance of the debate over the ontological status (personhood) of the fetus, and instead sought to shift the focus of the debate to the significantly disproportionate impact of pregnancy, birth, and parenting on women. There is also evidence, which we shall consider in Chapter 8, that judicial rulings in cases involving the right to refuse treatment reveal a tendency toward disparate treatment of remarkably similar statements by men and women who seek to decline life-sustaining measures.

Some feminist philosophers have also become advocates for communitarian ethics, which challenges the traditional atomistic liberal view of the self as abstracted from intimate and social relationships. The position one takes on such issues will influence the role that the family is allowed to have in health care decision making. The communitarian perspective in bioethics is most prominently featured in disputes about withholding or withdrawing life-sustaining treatment. Communitarian ethics attempts to shift the focus from respect for the individual autonomy of the patient to the familial, social, and cultural context in which the patient lives.[25] Ultimately, however, feminist bioethics, care ethics, and communitarian ethics seek to expand the range of factors that are taken into consideration in arriving at a conclusion about who should ultimately have the right to decide important questions about patient care.

1.12 Casuistry

The last of the approaches to ethics in health care that we will consider, one which also has experienced a revival in recent years, is the case-based reasoning approach known historically as casuistry. As will become clear when we shift our focus to law in the next section, casuistry parallels legal analysis in its reliance on paradigmatic cases. Like the law, the casuist approach maintains that rules and theories have no efficacy in isolation from their historical origins and evolution, as well as their prior application in similar cases (precedent).

Modern proponents of casuistry offer it as an antidote to the flaws they perceive in principalism, namely, its purported rejection of a role for cases in moral decision making other than by way of example or illustration. The casuist maintains that case analysis can and should be at the very core of moral reasoning, particularly in those situations in which pure theory, even enriched with derived principles and rules, does not provide an adequate basis for moral judgment.

Once again we find that the primary controversy is one of priority in bioethical decision making. Hardly anyone disputes either the existence of paradigmatic cases or that they can be instructive, by way of analogy, in other situations presenting similar facts and circumstances. The crux of the dispute is whether such cases speak for themselves without recourse to some undergirding or overarching moral theory or principle, or whether the case can only become significant and in some sense precedential by virtue of its exemplification of some theory or principle. This is where casuistry in bioethics and the case system of legal analysis part company. A single case at law, even an extremely significant one decided by the highest court in the jurisdiction, has no meaning by itself. In the words of one notable legal scholar: "Standing alone it gives you no guidance."[26] If that is true of formal, appellate legal decisions containing a definitive statement of the facts, the issue(s) presented, the relevant law(s), the disposition of the case and the reasons for that disposition, then how much more true will it be with regard to cases in bioethics? Indeed, what constitutes such a "case" lacks a definitive answer.

Beauchamp and Childress, in their discussion of casuistry in bioethics, refer to the Quinlan case, The Tuskegee Syphilis experiments, and the Quill case[27] as examples of paradigm cases offered by the proponents of casuistry as "sources of authority" or a "locus of moral certitude." With regard to Quinlan, what shall we say constitutes "the case"? Is it the official opinion of the New Jersey Supreme Court? But that is a statement of the case as a matter of medical jurisprudence, not medical ethics. Is it merely the medical facts of the case of Karen Quinlan – her medical history, family history, history of present illness, diagnosis, prognosis, proposed treatment, etc.? What if certain facts of that case are in dispute?

We encounter similar problems if Quill is offered as a paradigmatic case for casuistry in bioethics. Is the "case" the indictment of Dr. Timothy Quill by a grand jury for providing a lethal prescription with which his patient committed suicide and the resulting jury verdict of "not guilty"? Is it just the clinical facts of the patient's diagnosis, prognosis, regimen of care, and the prescription medication she used to end her life? Furthermore, what makes the case paradigmatic if we can have resort only to the facts?

Albert Jonsen, one of the most prominent advocates for casuistry in bioethics, maintains that the proper use of casuistry does not deny the relevance of principle and theory.[28] What varies from case to case, he suggests, is the extent to which the resolution of a case using the casuistic method will be driven by any particular theory or principle. However, inasmuch as reasonable minds may differ as to what theory or principle prevailed (or should prevail) in a particular paradigm case, they will differ on the nature of the precedent established by that case for future circumstances of a similar nature. If we consider the Quinlan case, it is possible to formulate arguments that it stands for a number of different theoretical propositions and principles, not all of which can be reconciled. Quill, *qua* physician, was certainly concerned about respecting the individual autonomy of his patient, who clearly had decisional capacity. But he also operated out of the clear conviction that beneficence did not dictate that Diane be denied the means to determine the time and manner of her own death. Yet critiques of Quill's conduct in the case of Diane can be launched from a deontological perspective – physicians must never become the causal agents of a patient's death, as well as a utilitarian perspective – physician-assisted suicide places us on a slippery slope from which we may then begin to justify voluntary active euthanasia.

The purpose of this section has not been to provide an exhaustive list of the theoretical approaches to bioethical decision making, or to weigh their relative strengths and weaknesses for the purpose of advancing one over the others, but rather to provide a brief introduction and overview of those which are most often discussed in the bioethics literature. Moreover, as we consider particular issues in bioethics and medical jurisprudence in subsequent chapters, it simply will not be possible to discuss each theoretical approach. However, on a very limited and selective basis we will attempt to see what light can be shed on particularly intractable controversies by one or more of these approaches.

2. THE LEGAL LANDSCAPE

In this section we will consider the four major sources of law in the legal system of the United States: the common law, constitutional law, legislation, and administrative law and regulation. In the next chapter we will briefly consider several theories of law as they relate to the debate over the proper relationship, if any, between law and morality.

2.1 The Common Law

The United States inherited a common law system of adjudication from England. The common law is essentially case-based. Disputes between parties about rights and responsibilities, be they in matters of contract, tort, descent and distribution, or some other substantive area of the law, are initially resolved at the trial level by a judge (bench trial) or a jury. The criminal law is essentially a dispute between "the people" of a particular jurisdiction and the alleged perpetrator(s) of a crime as to what in fact happened, as well as when, how, and involving whom. While the procedures and burden of proof differ significantly between the civil and criminal systems, both ultimately result, if not otherwise resolved by settlement (civil) or plea bargain (criminal), in a trial in which the entire proceedings are reduced to a verbatim transcript. The primary purpose of the transcript is to provide an accurate record upon which either or both sides of the case can seek a review of purported errors by a higher (appellate court).

Trial and appellate courts exist in both the state and the federal system, although most common law adjudication takes place at the state level. The federal system is reserved for disputes raising issues of federal law – usually statutory or constitutional – that meet the fairly strict requirements of federal jurisdiction. Both the state and federal court systems consist of intermediate appellate courts, in which the disposition of a trial court may be initially reviewed, and a court of last resort, usually known as the supreme court. On rare occasions, when major legal questions of first impression arise in the trial of a case, the intermediate court may be bypassed in order to obtain a prompt opinion of the highest court. This "fast-track" appellate review has characterized a number of prominent cases in bioethics.

The most salient feature of an appellate court review is that it consists exclusively of a review of the record of the proceedings at the trial level. There are no further evidentiary hearings, only a submission of the official record on appeal, written briefs submitted by the attorneys for the parties, and frequently (but not always), oral arguments before the court by the attorneys. Implicit in this fact about appellate review is that considerable deference is to be given to the factual findings of the trial court. Appellate courts generally will not overturn them unless there is simply nothing in the record to support them or they are otherwise clearly erroneous. However, as we shall see, in a number of prominent "right to die" cases appellate courts have flagrantly violated this principle of review. Appellate review ultimately results in a written opinion by the appellate court affirming in whole or in part or reversing the disposition of the case by the trial court. In addition to the opinion of the appellate court, there may be concurring and/or dissenting opinions by some of the judges. When a case is reversed, it is sent back to the trial court for a new trial or for other proceedings consistent with the ruling of the appellate court.

Opinions by the various levels of appellate courts in both the state and federal systems provide the precedents for which the common law is best known. It is important to note the distinction between binding precedent and what I shall refer to as "instructive precedent." When the New Jersey Supreme Court decided *Quinlan* in 1976, its holding became binding precedent on all state courts in New Jersey.[29] For courts in other jurisdictions, where it was often cited and discussed, it was "instructive precedent" only, i.e., such courts, if persuaded by its reasoning, might rule the same in similar cases so long as the prevailing law in their jurisdiction was not contrary, but they were under absolutely no obligation to do so for any precedential reasons.[30] A significant exception to this general rule is the U.S. Supreme Court. Its precedents are binding not only on all the federal circuit courts of appeal, but on all state courts as well, at least with regard to the Constitutional rights of citizens. Thus, for example, after the *Roe v. Wade* decision in 1973, no court in the United States could determine that a pregnant woman in the first trimester of pregnancy did not have a Constitutional right to privacy that included the right to seek an abortion.

The development and application of the common law bears the closest resemblance to the method of casuistry in bioethics advocated by Jonsen and others. When a particular common law decision establishes an important legal precedent, e.g., that competent patients have the right to refuse life-sustaining medical interventions unless one or more state interests can be deemed so compelling as to limit the exercise of that right, then it becomes a paradigm case by which to analyze the facts of future cases. In order to rebut the presumption that a patient may refuse any medical procedure, the party with the burden of persuasion must show that the facts in the present case are significantly different from those in the case which established the precedent such that a different outcome is warranted. As noted earlier, common law precedent-establishing (paradigm) cases can be applied more consistently and effectively because of the format that appellate decisions follow. That format includes an articulation of the facts of the case as established by the trier of fact at the trial court level, the legal issues presented and resolved, the precise holding of the case, and the legal reasoning that supports the holding. There is no comparably definitive and detailed version of the paradigm cases relied upon by the casuists.

2.2 Constitutional Law

The Constitution of the United States is, fundamentally, the charter of our particular form of government. It establishes the role and responsibility of the federal government and its relationship to the states and to what is more generically referred to as "the people." Much of the Constitution sets forth in fairly precise and hence unambiguous language the particulars of the federal government, e.g., the authority of the houses of Congress and how their members shall be selected, as well as the nature of the executive and judicial branches. The Constitution also provides for the process for amending that charter. The first ten amendments, known as the Bill of Rights (ratified in 1791), and the Fourteenth Amendment (ratified 1868) provide much of the basis for constitutional adjudication in areas of significance to bioethics.

It is an important fact of our constitutional history that the need for and the wisdom of a Bill of Rights was hotly contested among the delegates to the Constitutional Convention. Among the concerns repeatedly expressed was that by specifically listing rights which the people retained as against the federal government, there might be an inference that those rights not enumerated had been somehow waived or forfeited. When the Bill of Rights was ratified, this concern was addressed by the Ninth Amendment, which provides: "The enumeration in the Constitution, of certain rights, shall not be construed to deny or disparage others retained by the people," and by the Tenth Amendment, which provides: "The powers not delegated to the United States by the Constitution, nor prohibited by it to the States, are reserved to the States respectively, or to the people." A considerable literature has developed as a result of the failure of the Ninth Amendment to play any significant role in American constitutional jurisprudence.[31] We will briefly consider one of the few exceptions in Chapter 6 when we discuss Justice Goldberg's concurring opinion in *Griswold v. Connecticut*.

The lingering question, not finally resolved until the ratification of the Fourteenth Amendment and the subsequent decisions of the Supreme Court interpreting and applying it, was whether the people of the United States were protected by the Bill of Rights not only from infringement of their liberty by the federal government, but also by the states. The most important passage of the Fourteenth Amendment in this regard is the following portion of Section 1: "No State shall make or enforce any law which shall abridge the privileges or immunities of citizens of the United States; nor shall any State deprive any person of life, liberty, or property, without due process of law; nor deny to any person within its jurisdiction the equal protection of the law."

In marked contrast to much of the unambiguous language of the articles of the Constitution, the amendments use terminology that is open to differing and hence potentially conflicting interpretation. For example, the privileges and immunities of citizens of the United States are nowhere specifically enumerated. Moreover, the Supreme Court has been exceedingly parsimonious in recognizing them in the course of Constitutional adjudication.[32]

Perhaps the most contentious aspect of constitutional interpretation, but one which also has particular significance for bioethical issues, is that of substantive due process. The passage from the Fourteenth Amendment quoted above appears straightforwardly to provide only that a state may not deprive a person of life, liberty or property *without due process of law*. Once the requisites of due process have been established, a state might seem to be recognized as having the power to deprive a person of life, liberty or property so long as the procedural niceties are scrupulously followed. However, in a series of cases between 1890 and 1937, the Supreme Court scrutinized numerous state laws against what clearly emerged as a substantive and not merely a procedural conception of the protection of life, liberty and property afforded by Fourteenth Amendment against infringement by the state. Substantive due process was most essentially, for purposes of this analysis, the recognition of a sphere of individual liberty (as we shall see in Chapter 6, an uneasy blend of autonomy and privacy) protected from unjustifiable intrusion by government. The following language by Justice Harlan effectively articulates the contemporary understanding of substantive due process:

> The full scope of the liberty guaranteed by the Due Process Clause cannot be found in or limited by the precise terms of the specific guarantees elsewhere provided in the Constitution. This "liberty" is not a series of isolated points picked out in terms of the taking of property; the freedom of speech, press, and religion; the right to keep and bear arms; the freedom from unreasonable searches and seizures; and so on. It is a rational continuum which, broadly speaking, includes a freedom from all substantial arbitrary impositions and purposeless restraints . . . and which also recognizes . . . that certain interests require particularly careful scrutiny of the state needs asserted to justify their abridgement.[33]

What Justice Harlan describes is a realm of unenumerated rights too numerous to be mentioned specifically in the Bill of Rights but envisioned by it in the Ninth Amendment's careful distinction between those rights "enumerated" in the Constitution and the first eight amendments and "others retained by the people." In the first half of the Twentieth Century, the Supreme Court had occasion to identify and justify a few of these unenumerated rights that come within the Constitution's fulsome concept of liberty. Among the unenumerated rights recognized by the Court were the right of the parent to teach a child a foreign language[34] and to send a child to a private school,[35] the right to procreate,[36] and the right to be free of certain bodily intrusions.[37] In Chapters 6 and 7 we will consider in some detail the Court's two most far-reaching substantive due process cases, *Griswold v. Connecticut* and *Roe v. Wade*, which directly interjected constitutional jurisprudence into the newly-recognized realm of biomedical ethics.

2.3 Legislation

The sources of legislation are the legislatures of each state and the Congress of the United States. When a citizen cries out in frustration "there ought to be a law," more often than not their entreaty is made to an appropriate legislative body. Such bodies, comprised as they are by elected officials, have not been unresponsive to these cries. As a result, some critics would say, American society is drowning in (or straightjacketed by, depending upon your preferred metaphor) a plethora of laws requiring this and forbidding that which is unprecedented in the history of Western democratic government.

As we consider the common law and constitutional cases raising important issues in bioethics, we will often encounter a proclivity of judges, especially those of a conservative philosophical stance, to relegate issues which raise strong and conflicting ethical positions to the legislature. The physician-assisted suicide cases received precisely this sort of treatment by the justices of the Supreme Court. In delivering the Opinion of the Court, Chief Justice Rehnquist noted in conclusion: "Throughout the Nation, Americans are engaged in an earnest and profound debate about the morality, legality, and practicality of physician-assisted suicide. Our holding permits this debate to continue, as it should in a democratic society."[33] In his concurring opinion, Justice Souter was more direct:

> Legislatures . . . have superior opportunities to obtain the facts necessary for a judgment about the present controversy. Not only do they have more flexible mechanisms for fact finding than the Judiciary, but their mechanisms include the power to experiment, moving forward and pulling back as facts emerge within their own jurisdictions. There is, indeed, good reason to suppose that in the absence of a judgment for respondents here, just such experimentation will be attempted in some of the states.[39]

What Justice Souter's panegyric on the legislative process overlooks is that the only states in which physician-assisted suicide legislation has been seriously considered or, in the case of Oregon, enacted, have been those which permit voter-initiated referenda. Indeed, such referenda led to a law permitting (in compliance with a strict regulatory process) physician-assisted suicide in Oregon which the legislature then sought to undo by forcing a second vote on the proposition through a rarely used legislative prerogative. Similarly, physician-assisted suicide referenda were narrowly defeated in Washington and California. The vast majority of state legislatures in which a bill was introduced by an elected representative have routed the bill to a hostile committee where it has died a quick and premature death. This legislative antipathy toward grappling with the issue of physician-assisted suicide stands in stark contrast to a number of recent public opinion polls indicating that a majority of Americans support the legalization and regulation of the practice of physician-assisted suicide for competent, terminally ill patients.

While it is true that in theory legislatures are deliberative bodies whose rules and procedures enable extensive consideration of a wide range of viewpoints, and the weighing of the positive and negative implications of proposed legislation, it is also true that some of the most contentious bioethical issues have become the political footballs of factions and interest groups that make a mockery of rational and dispassionate legislative analysis. Reliable public opinion polls may reveal that a majority of the citizens of a state support some particular position on an issue, e.g., legalization and regulation of physician-assisted suicide or allowing malpractice suits against managed care organizations, yet a majority of the members of the legislature may oppose the majority view of the citizenry and vote accordingly.

The well-worn piece of folk wisdom from the legislative trenches is that there are two things one should never closely observe the production of – sausage and legislation. The rationale is that while the finished product may be presentable and even digestible, the process itself, and the basic ingredients in both instances, are often quite unpalatable. The legislative process can, as a practical matter, work both ways. A bill as originally introduced may be clearly conceived, targeted to achieve particular objectives, and carefully crafted to minimize ambiguities and unintended consequences. However, in the process of committee review, floor debate, friendly and hostile amendments, and the bargaining and compromise necessary to achieve a favorable vote, something quite different and much less effective may in fact become law. On the other hand, a bill as introduced might be ill-conceived, hopelessly ambiguous, and/or fraught with unintended consequences. Yet, rather than simply being quickly voted down in a committee, it may in fact be rehabilitated through the very same process and emerge to enactment as a salutary piece of legislation.

There is also the problematic phenomenon, as we shall see in *Griswold*, of legislation as symbol or window-dressing. In the case of *Griswold,* the State of Connecticut had an "ancient" statute making it a criminal offense to use contraceptives, as well as a general accessory statute which made it a criminal offense to aid another in committing an offense. Using the two statutes in tandem, the state could in theory prosecute a physician who supplied a patient with the means for contraception, as well as the patient who used the contraceptive. Nothing in the language of the statute made any distinction between married couples or unmarried persons; all usage was proscribed. As a matter of fact, no one had ever been prosecuted for violating the statute, nor had any prosecutors even intimated that they would be inclined to do so under certain circumstances. Nevertheless, despite increasingly vocal opposition to the statute, the Connecticut legislature demonstrated no inclination whatsoever to repeal the law.

The supposed rationale for such laws – never enforced and never repealed – is that they tacitly affirm certain moral principles without extracting a price for those whose conduct violates those principles. In the case of the Connecticut statute, the principle was actually theological rather than one of secular morality – the opposition of the Roman Catholic Church to contraception. If in fact it could be shown that the sole or even the primary reason why a majority of the Connecticut legislature declined to repeal the statute was fear of offending the Catholic voters of the state, even in the face of a uniform open and notorious refusal on the part of prosecutors to enforce the law, then a strong argument can be formulated that such legislation is pernicious. First, it does not even represent a legitimate attempt to enforce a common morality (assuming *arguendo* such an attempt can ever be legitimate, as we shall discuss at length in Chapter 3), since a majority of the citizens of Connecticut, at least in the mid-1960's, did not believe that the use of contraceptive devices, at least by married persons, was immoral. Second, such never-to-be-enforced statutes can contribute to a disrespect for law in general. It is one thing to exercise prosecutorial discretion, so that criminal statutes may be enforced selectively (albeit not discriminatorily), but blanket nonenforcement is a very different thing from a legitimate decision not to prosecute made on a case-by-case basis.

A concluding point for this preliminary discussion of legislation is that it is always subject to interpretation (or misinterpretation) or even nullification by the courts. There may also be a kind of point-counterpoint in which legislatures and courts, as well as administrative agencies, which we will consider next, respond to actions by one another. For example, in recent years it has been increasingly recognized that state medical licensing boards have had a negative impact on the quality of pain relief provided by physicians because of their (arguably) overzealous investigation and discipline of physicians for allegedly over-prescribing opioid analgesics. In response, a number of state legislatures have enacted "intractable pain" statutes which preclude such boards from taking disciplinary action against physicians "solely" because they have prescribed opioid analgesics for chronic pain.[40]

When the legislature is displeased with a court's interpretation of a statute, it can amend it so as to make clearer its intention. For example, when the North Carolina Supreme Court upheld disciplinary action against a physician who had provided selected patients with homeopathic remedies pursuant to their fully informed consent,[41] the North Carolina legislature amended the statute specifying grounds for disciplinary action so as to preclude the licensing board from disciplining a physician for providing non-traditional therapies unless the board produces competent evidence that the treatment poses a greater risk than the prevailing practice or that the alternative treatment is generally ineffective.

2.4 Administrative Agency Regulation

In the last portion of the previous section we considered certain legislative and judicial responses to the exercise of authority by state medical licensing boards, which are examples of administrative agencies. Such licensing boards, which exist in each state to regulate a wide variety of professions and occupations, are generally granted a great deal of discretionary authority by both courts and legislatures because they are presumed to possess expertise in the regulated activity that far exceeds that which exists in courts and legislatures. They are charged with conducting their affairs in the public interest. When one of their actions is legally challenged, a court will not usually overturn it absent a showing that it was arbitrary, capricious, not supported by competent and credible evidence, or in excess of its statutory authority.

In the health care setting, much administrative action is at the state and local level. However, Congress has stepped in and created federal remedies for problems that are nationwide, or in instances when the states have failed or refused to institute appropriate remedial actions. One important example is the federal Emergency Medical Treatment and Active Labor Act (EMTALA),[42] referred to colloquially as the "anti-dumping" statute. The widespread evil Congress sought to address in 1986 with this legislation was the transfer from one hospital emergency room to another of patients who were medically unstable or pregnant women in active labor for economic reasons, i.e., patients without health care insurance. While many EMTALA actions are initiated by the patients who claim to be the victims of improper transfers, and are often combined with common law medical malpractice claims, the Health Care Financing Administration (HCFA), which promulgates implementing regulations for EMTALA, has the authority to terminate a hospital's Medicare Provider Agreement if it finds a violation of the Act. The Office of Inspector General (OIG) may impose civil monetary penalties on hospitals or physicians up to a maximum of $50,000 per violation.

A complete discussion and analysis of EMTALA would take us well beyond the scope of this chapter. What is significant about EMTALA for our purposes is that it exemplifies a federal legislative and administrative response to a pervasive problem in the health care delivery system, one which has not only clinical and economic but also ethical implications. Indeed, there seems to have been an expectation, or at least the hope (perhaps naive) on the part of Congress and HCFA that hospitals charged with EMTALA violations and subject to public notice of their conduct would quickly reform their practices and settle the claims out of court. That has not turned out to be the case.[43] The patients who are most commonly victimized by violations of EMTALA have virtually no economic or political power, and patients of higher socio-economic status tend not to identify with the poor and underserved in ways that might motivate a concerted refusal to patronize hospitals with a reputation for dumping indigent patients.

Another example of federal intervention in the health care setting is the Patient Self Determination Act (PSDA), enacted in 1990.[44] Like EMTALA, the PSDA is linked to Medicare providers and empowers HCFA to impose on institutions which violate the act the draconian sanction of loss of all Medicare and Medicaid funding. Enacted in the wake of the publicity surrounding the case of Nancy Cruzan and its review by the U.S. Supreme Court, the PSDA requires all health care institutions which are Medicare or Medicaid recipients to provide each patient, or the patient's surrogate, upon admission or as soon thereafter as practicable, with written information about an individual's rights under State law (statutory or common) to make decisions about medical care, including the right to refuse treatment and to execute advance directives. The institution is also required to provide each patient with the written policies of the institution with respect to the implementation of those rights.

It is important to note that the PSDA creates no substantive right to refuse treatment or to execute advance directives, it merely imposes a duty upon the institution to inform each patient of their rights under state law and of institutional policies, if any, affecting those rights. While the expectation of Congress was that the PSDA would increase the use of advance directives by expanding the knowledge of patients about their rights to make treatment decisions and to engage in advance care planning through the use of written advance directives, those expectations have not been fulfilled. Studies continue to reveal that only about 10% of adults in the United States have executed a written advance directive,[45] and that in some of the most prestigious academic medical centers in the country even concerted efforts to improve conversation between physicians and patients and their families about wishes and goals in critical care situations have failed miserably.[46] These issues will be considered at length in Chapter 5.

Some of the most influential federal statutes and administrative regulations in the health care field are those governing fraud and abuse in the Medicare and Medicaid reimbursement systems. The penalties for conduct determined to constitute fraud and abuse, such as submitting a false claim for reimbursement or offering or receiving unlawful kickbacks, rebates or bribes, can be draconian. For example, the Social Security Act makes it a felony to knowingly make any false statement of a material fact in any application for payment under a federal healthcare program which may be punished by fines of up to $25,000 as well as up to five years imprisonment.[47] The Federal False Claims Act provides for civil penalties for knowingly false claims of between $5,000 and $10,000 per violation plus three times the amount of damages sustained by the government.[48] The Office of Inspector General (OIG) of the Department of Health and Human Services (HHS) has the authority to exclude from participation in the Medicare program individuals or entities convicted of any felony related to health care fraud or abuse, which can in turn lead to disciplinary action by state medical licensing boards.

The entire field of Medicare and Medicaid fraud and abuse investigation and enforcement has produced a tremendous growth in health law because the stakes are so high for those providers whose practices have come under administrative scrutiny. However, despite the "ethical" implications of fraudulent behavior on the part of health care providers, bioethics has shown little interest in this subject. Unlike many other legal issues in health care, there has been little or no convergence between health law and bioethics with regard to the Medicare and Medicaid systems.

2.5 Overview

From this very basic review of the key concepts, principles, and categories in bioethics and law, it is possible to discern certain consistent themes as well as a few significant asymmetries. Certainly at the present time both bioethics (essentially irrespective of the particular theory one espouses) and law offer strong support for respecting patient autonomy and acting for the benefit of the patient when providing medical treatment. But as we shall see in Chapter 4, the long history of medical paternalism was supported, not challenged, by traditional medical ethics. Indeed, one of the salient features of the transition from a medical ethics as the exclusive province of the medical profession to a bioethics that is genuinely multidisciplinary is the newfound emphasis upon respect for patient autonomy as an ethical principle.

While traditional medical ethics consistently acknowledged that medicine is a moral undertaking with significant ethical responsibilities running from physician to patient, the tenor of these responsibilities, viewed from a legal perspective, bore a much closer resemblance to those of a parent for a child than those of a fiduciary to a fully responsible adult who has entrusted certain important matters to the expertise of another. If there is a parallel between traditional medical ethics and the law, it is the tendency, at least before the middle of this century, for the law (in all of its varieties) to follow in general a hands-off philosophy with regard to the practice of medicine. Only the rare, egregious situations would find their way into the courtroom or the legislative assembly. Now, all of that has changed with a vengeance. But before we begin to examine in detail the nature and significance of some of these changes, it is important to consider the historical tension between law and morality. We do this because ultimately a question which must be addressed is whether the law's significant involvement in matters of bioethics makes sense and more particularly whether it has legitimacy and efficacy. To some extent our examination of these larger questions will be influenced by traditional views of the proper or necessary relationship, or lack thereof, between law and morality.

ENDNOTES AND REFERENCES

1. Tom L. Beauchamp and James F. Childress, *Principles of Biomedical Ethics*, Fourth Edition (New York: Oxford University Press, 1994), p. 189.
2. Hippocrates, *Decorum* (2), William Jones, trans. (Cambridge: Harvard University Press, 1967), p. 297.
3. Interestingly, one extensive commentary on Percival's *Medical Ethics* argues that it is misnamed, since the work deals almost exclusively with what might more aptly be characterized as medical etiquette, which concerns the physician's responsibilities toward other physicians, not toward patients. Chauncey D. Leak, ed., *Percival's Medical Ethics*

(Baltimore: Williams & Wilkins Co. 1927).
4. American Medical Association Code of Medical Ethics, (N.Y.: William Wood & Co., 2nd ed., 1868).
5. 63 Federal Register 16296.
6. In a 1983 report, the President's Commission for the Study of Ethical Problems in Medicine and Biomedical and Behavioral Research issued a report, with scholarly and detailed appendices, entitled *Securing Access to Health Care.* The Commission elected not to propose any new health policy initiatives to address the well-documented problems of access to care, but rather to provide a framework within which debates about health policy might take place. Nevertheless, the Commission did conclude that "society has an ethical obligation to ensure equitable access to health care for all." Thus far, the United States continues to fail to fulfill that obligation to increasing numbers of its citizens each year.
7. Beauchamp and Childress, note 1, pp. 37-40.
8. Beauchamp and Childress, note 1, p. 111.
9. Olmstead v. United States, 277 U.S. 438, 478 (1928) (Brandeis, J., dissenting).
10. For a detailed discussion of utilitarianism see David Lyons, *Forms and Limits of Utilitarianism* (Oxford: Oxford University Press, 1965).
11. For a general discussion of Kant's moral philosophy see H. J. Paton, *The Categorical Imperative* (Chicago: University of Chicago Press, 1948).
12. Immanuel Kant, *Fundamental Principles of the Metaphysics of Morals*, § 1, Thomas Kingsmill Abbot, trans., (London: Longmans Green, 1927).
13. Alasdair MacIntyre, *After Virtue – A Study in Moral Theory*, 2nd ed., (Notre Dame: University of Notre Dame Press, 1984).
14. MacIntyre, note 13, p. 191.
15. MacIntyre, note 13, p. 195.
16. For further discussion of the powerful influence of health care institutions on the virtues in medical practice see Leon Kass, "Practicing Ethics: Where's the Action?" *Hastings Center Report* (1990), 20: 5-12.
17. Earl E. Shelp, ed., *Virtue and Medicine* (Dordrecht: Reidel, 1985).
18. Edmund D. Pellegrino and David C. Thomasma, *The Virtues in Medical Practice* (New York: Oxford University Press, 1993).
19. MacIntyre, note 13, p. 220.
20. MacIntyre, note 13, p. 218.
21. Kant was neither alone in nor the originator of this point of view. It was first asserted by Aristotle, and further developed by Rousseau and Hegel.
22. Carol Gilligan, *In A Different Voice* (Cambridge: Harvard University Press, 1982).
23. For a more general discussion of the ethics of care *see* Nel Noddings, *Caring: A Feminine Approach to Ethics and Moral Education* (Berkeley: University of California Press, 1984).
24. *See, e.g.,* Sara T. Fry, "The Role of Caring in a Theory of Nursing Ethics," in Helen Baccarat Holmes and Laura M. Pard, eds., *Feminist Perspectives in Medical Ethics* (Bloomington: Indiana University Press, 1992).
25. John Hardwig, "What About the Family?," Hastings Center Report, (1990), 19(2): 5-10.
26. Karl Llewellyn, *The Bramble Bush – On Our Law and Its Study* (New York: Oceana Publications, Inc., 1950), p. 48.
27. The case to which they refer is not the physician-assisted suicide case, *Quill v. Vacco*, but the same Dr. Quill's chronicle of his provision of a lethal prescription to his terminal patient Diane. Timothy E. Quill, *Death and Dignity: Making Choices and Taking Charge* (New York: Norton, 1993), Ch. 8.
28. Albert R. Jonsen and Stephen E. Toulmin, *The Abuse of Casuistry: A History of Moral Reasoning* (Berkeley: University of California Press, 1988), p. 10.
29. In the Matter of Karen Quinlan, 355 A.2d 647 (N.J. 1976).

30. One of the more remarkable examples was provided by the Missouri Supreme Court in the Cruzan case. After an extensive review of factually similar cases in other jurisdictions, in which there was a clearly distinguishable trend toward a "best interests" approach when an incompetent patient had failed to leave clear written directives, the majority opinion declined to follow this trend and declared that absent clear and convincing evidence that the incompetent patient would wish to have life support withdrawn, such support must be provided as a means of demonstrating the reverence that the citizens of Missouri have for all human life, regardless of its quality, even though continued life support might not be in the best interests of the patient. The Missouri court is also one of those that failed to accord the findings of fact by the trial court their traditional presumption of validity. Cruzan v. Harmon, 760 S.W.2d 408 (Mo. Banc 1988).
31. *See, e.g.*, Bennett B. Patterson, *The Forgotten Ninth Amendment* (Indianapolis: Bobbs Merrill, 1955); Luis Kutner, "The Neglected Ninth Amendment: The Other Rights Retained by the People," *Marquette L. Rev.* (1968), 51: 121-142.
32. In Edwards v. California, 314 U.S. 160 (1841) Justice Jackson stated in a concurring opinion that "instances of valid 'privileges or immunities' must be but few" (p. 183).
33. Poe v. Ullman, 367 U.S. 497, 543 (1961) (Harlan, J., dissenting).
34. Meyer v. Nebraska, 262 U.S. 390 (1923).
35. Pierce v. Society of Sisters, 268 U.S. 510 (1925).
36. Skinner v. Oklahoma, 316 U.S. 535 (1942).
37. Rochin v. California, 342 U.S. 165 (1952).
38. Washington v. Glucksberg, 138 L.Ed.2d 772, 797 (1997).
39. 138 L.Ed.2d 771, 830 (1997) (Souter, J. concurring).
40. See, e.g., Colo. Rev. Stat. §12-36-117(1.5)(a) (1997).
41. In Re Guess, 393 S.E.2d 833 (Sup. Ct. N.C. 1990).
42. 42 U.S.C.A. §1395dd.
43. Barry R. Furrow, "An Overview and Analysis of the Impact of the Emergency Medical Treatment and Active Labor Act," *Journal of Legal Medicine* (1995), 16: 325-355, pp. 352-353.
44. 42 U.S.C.A. §1395 cc.
45. Nancy C. Elder, F. David Schneider, Steven C. Zweig, Philip G. Peters, John W. Ely, "Community Attitudes and Knowledge about Advance Care Directives," *Journal of the American Board of Family Practice* (1992), 5: 565-572.
46. The SUPPORT Principal Investigators, "A Controlled Trial to Improve Care for Seriously Ill Hospitalized Patients," *Journal of the American Medical Association* (1995), 274: 1591-1598.
47. Social Security Act, §§1128B(a); 42 U.S.C.A. §1320a-7b.
48. 31 U.S.C.A. §3729.

Chapter 3

A HISTORICAL PERSPECTIVE ON THE RELATIONSHIP BETWEEN LAW AND MORALITY

1. THE DEBATE BETWEEN NATURAL LAW THEORY AND LEGAL POSITIVISM

Two theories which have dominated Anglo-American jurisprudence in the modern era incorporate opposing positions on the issue of the necessary or proper relationship between law and morality – natural law and legal positivism. We will begin with natural law because it has the longest lineage.

2. NATURAL LAW THEORY

Natural law theory can be traced as far back into history as ancient Greece, with strains of the theory appearing in Plato's *Laws* and the *Republic* and Aristotle's *Politics* and the *Nichomachean Ethics*. While later development, particularly that of Aquinas, has given natural law theory a religious aura, a natural law perspective does not necessarily presuppose any religious viewpoint. What it does posit is a natural moral order that is present in the universe and accessible to human reason. This natural moral order is uniform across time and place, and provides the basis upon which to evaluate positive law, i.e., the laws embodied in legislation, constitutions, and judicial decisions. Whether a particular instance of positive law – a statute or court ruling – is consistent with justice or the freedom and dignity to which human beings are entitled *qua* persons is a moral assessment that a natural law perspective enables one legitimately to make.

Thus, the meaning of "natural" in a natural law theory is the natural moral order of which human beings are a part and to which they have unique access by virtue of their mental faculties. Natural law theory maintains that there is an essential relationship between law and morality, and positive laws which are inconsistent with the moral truths of the natural moral order fail to partake of the character of law. When we come to the Enlightenment period, and more particularly the writings of Grotius and Locke, we are introduced to the equally important concept of natural rights. The Lockean account of natural rights which precede and supersede particular governments was embraced by many, if not most, of those political leaders whose deliberations brought forth the Declaration of Independence and the Constitution of the United States. As alluded to in Chapter 2, and as will be more fully developed in Chapter 6, the substantive rights to liberty and privacy that have been found to be implicit in the Due Process Clause of the Fourteenth Amendment draw heavily upon the Lockean concept of natural rights.

Some commentators have argued for a clear line of demarcation between ancient and medieval approaches to natural law and contemporary natural law theory.[1] According to this account, contemporary natural law theory should be understood as a response to legal positivism, which we will consider next. Perhaps the most notable contemporary exponent of a naturalistic legal theory is Ronald Dworkin, whose recent work on bioethical issues we will consider in Chapters 7 and 8.

Dworkin maintains that any legal system worthy of the term will necessarily include legal principles which are, in fact, a type of moral proposition.[2] Thus, the conceptual and pragmatic distinction between principles of law and principles of morality is a false and misleading one, for moral evaluation is an essential element in both the description and the understanding of law. This is particularly true when courts interpret and apply provisions of the Constitution to cases. It gives rise to what Dworkin calls a "moral reading of the Constitution," and which he insists is justifiable upon the ground that "the Bill of Rights can only be understood as a set of moral principles."[3] From such a naturalistic perspective on law and legal systems, suggestions that the Supreme Court should decline to rule in cases raising contentious moral issues, or that they should rule only on narrow legal points which leave the moral issues open and undecided, is not only incorrect but incoherent.[4] In Chapter 9 we will consider the views of one of Dworkin's harshest and most outspoken critics, the prolific federal chief judge of the U.S. Court of Appeals for the Seventh Circuit, Richard Posner. His attack on Dworkin and other moral philosophers is based upon the contention that moral philosophy (and, by implication, bioethics) is not helpful in deciding legal cases, even those that raise profound moral questions.

3. LEGAL POSITIVISM

The positivist position is that while there may in fact be a great deal of overlap or common ground between a society's laws and its morality, there is no necessary connection between them. It is an open question, subject to legitimate political debate, whether it is better for a society to have considerable or negligible congruence between its legal rules and its moral principles. Positivism maintains that legal rules derive their validity from the process through which they are produced, not from their moral value. Hence, a law can fail every conceivable "moral" test and yet be a valid and enforceable law because it comports with a society's established criteria for generating and recognizing a law.

Legal positivists do not present a monolithic or univocal theory. Two notable and disparate views are offered by John Austin and H.L.A. Hart. For Austin, the quintessential characterization of law is that of an order by the sovereign backed-up by the threat of sanction in the event of noncompliance.[5] The authority of law is derived from its source; it is a matter of fact, not value. The authority of law is also related to its enforceability. Obviously, a law which does not, in fact, constitute a command of the sovereign, will pose enforcement problems because the machinery of law enforcement – criminal or civil – will be difficult if not impossible to mobilize under such circumstances. Austin's formulation also suggests a distinction between law and morality which we will consider further in the concluding chapter, which has to do with the consequences for violating prevailing moral standards in a society, particularly professional codes of conduct for those in health care.

H.L.A. Hart's version of legal positivism differs markedly from that of Austin, at least if a narrow and literal interpretation of the latter is applied. For example, Hart notes that the sovereign in most Western nations is not so much a particular person as an office or a governmental body, which maintains its authority without interruption as the holders or members change over time. Also, there are many types of laws which cannot properly be conceived of as commands, but rather as empowerments – either of governmental officials or, in the case of laws recognizing or conferring individual rights, citizens. Finally, and most interestingly, Hart insists that in order to fully understand the authority of law, we need an explanation of how law can and does motivate general compliance in the absence of sanctions. The need to invoke sanctions in order to generate compliance with the law, he suggests, constitutes strong evidence that a general recognition of the authority of law is absent.[6] Sanctions, on this account, provide a reassurance to the majority of citizens who have cultivated "the habit of obedience" to law, that they will not be disadvantaged through their obedience by the minority who have not. On this view, people obey laws not because of any natural inclination, but as a result of an informal risk/benefit analysis. What we will later consider in the context of particular issues in bioethics is whether moral suasion, operating in the absence of legal sanctions, is capable of consistently generating "the habit of obedience."

One of the canonical iterations of the dangers inherent in confusing law and morality can be found in an address delivered by Oliver Wendell Holmes in 1897.[7] Holmes maintains that while "the law is the witness and external deposit of our moral life" and the history of the law "is the history of the moral development of the race," nevertheless if one is to understand the law, i.e., to know the law in its most fundamental sense, one must "look at it as a bad man, who cares only for the material consequences which such knowledge enables him to predict, not as a good one, who finds his reasons for conduct, whether inside the law or outside of it, in the vaguer sanctions of conscience."[8] Holmes' "bad man" is the quintessential amoral person who complies with the norms of society if, and only if, the sanctions for noncompliance impose burdens that are greater than the personal benefits to be derived through the transgression. Moreover, the law on Holmes' view cares not one whit about a person's motives for compliance. Whether a person is law-abiding because the law comports with his or her sense of what is right from a moral standpoint, or merely because the cost of violating the law is too high, the result is the same. We will return to Holmes' views in Chapter 9 in the context of our discussion of the challenge by Posner to the relevance of moral philosophy to judicial decision making.

The competing views of the relationship (or lack thereof) between law and morality were brought into stark focus during the Nuremberg trials of Nazi war criminals immediately following the end of World War II. The Charter of the Tribunal granted jurisdiction over crimes against peace, war crimes, and crimes against humanity. Crimes against humanity included murder, extermination, enslavement, deportation, and "other inhumane acts committed against any civilian population before or during the war."[9] The defendants challenged their prosecution on classical legal positivist theory, i.e., their alleged criminal actions were undertaken pursuant to the orders of superiors of a sovereign nation. Therefore, they should be immune from legal liability and their actions and the consequences of those actions should be imputed to the state. There can be no higher authority upon which to challenge acts perpetrated pursuant to the commands of a sovereign state.

The response of the prosecution, and ultimately the Tribunal, relied heavily upon the essential elements of natural law theory. For example, Sir Hartley Shawcross, chief prosecutor for the United Kingdom, responded to the defense position in the following terms:

> No rule of international law . . . provides immunity for those who obey orders – which whether legal or not in the country where they are issued – are manifestly contrary to the very law of nature from which international law has grown and . . . of common conscience and elementary humanity.[10]

To say this is not necessarily to contend that because many of the laws of the Nazi era in Germany constituted crimes against humanity, they were not laws at all in any meaningful sense. Rather, as Hart points out, "if . . . we speak plainly, we say that the laws may be law but too evil to be obeyed."[11] Nevertheless, critics of even Hart's less rigid and austere brand of positivism suggest that it fails to account in any intelligible way for what Lon Fuller has characterized as "the moral obligation of fidelity to law." The authority to make law, Fuller maintains, "must be supported by moral attitudes that accord to the competency it claims . . . a morality external to law, which makes law possible."[12]

As we consider some of the important legal decisions in the field of bioethics, it should become quite clear that most of them exude a naturalistic rather than a positivistic aura. At the same time, many courts take the position that they are deciding legal issues based upon legal principles rather than making moral pronouncements. Only through a case-by-case analysis can one determine the manner and extent to which extra-legal moral sentiments are influencing the outcome. The appropriateness of introducing such sentiments will be a major consideration in Chapter 9.

4. THE LEGAL ENFORCEMENT OF MORALITY

The perennial debate over the feasibility and the propriety of efforts to compel moral behavior or prevent immoral behavior offers yet another perspective on the inter-relationship between law and morality. The debate focuses not on the obviously overlapping cases in which legal, moral, and even religious precepts sanction certain types of conduct, such as murder, assault and battery, or child abuse, but on activities in which the resulting harm is not merely subtle, but arguably non-existent except to those with exquisite sensibilities. In his seminal four volume work, *The Moral Limits of the Criminal Law*, philosopher Joel Feinberg considers the question: "What sorts of conduct may the state rightly make criminal?" Two of the four volumes, *Offense to Others* and *Harmless Wrongdoing*, address those activities whose prohibition through the criminal law are most controversial and most likely to constitute illegitimate efforts to legally enforce morality. Feinberg uses the term "offense" in contradistinction to the word "harm," to denote a negative reaction that does not rise to the level of an actual setback to the interest of another in either the manner or the degree of a harm. Among those that he considers are desecration of sacred symbols, brandishment of odious symbols such as swastikas or the hooded garb of the Ku Klux Klan, and obscenity and pornography. *Harmless Wrongdoing* considers as prime candidates for the legal enforcement of morality "free-floating, nongrievance evils" such as the violation of social taboos or conventional immoralities on the order of extramarital and homosexual intercourse. At one time or another in the history of the United States, many such activities have been prohibited by law. For example, the sodomy laws of some states have made it a crime to engage in certain types of sexual activities, even in complete privacy and by consenting adults of the opposite sex, including married couples. While enforcement of such laws may have been problematic both in theory and practice, that fact did not appear to impede their enactment by the legislature. A prime example of such a law will be the focus of our discussion in Chapter 6.

The most articulate and zealous advocate for such legislation – what Feinberg refers to as "legal moralism" – was Sir Patrick Devlin. His views on the propriety of, indeed the necessity for, the legal enforcement of morality were set forth in his book *The Enforcement of Morals*.[13] According to Devlin, a common and consistent moral fabric is essential, not merely to the flourishing of a society, but to its very survival. This claim, known as the "social disintegration thesis," is one for which very little empirical evidence exists. Actually, Devlin's claim is an even stronger one, as Feinberg rightly points out, since he maintained that any change in the norms of a society was necessarily a change for the worse. Inherent in Devlin's view is a concept of society characterized by rigidity and fragility. It is rigid in the sense that a society whose norms change over time, perhaps because the legal framework has allowed differing moral views to gain a foothold and establish a common currency, is deemed not to have evolved, but to have taken on a new identity. It is fragile because it postulates that any moral views that are at variance with the original moral stance of the society pose a significant threat to the continuing existence of that society.

Devlin's perspective on the relationship between the norms of a society and its continuing existence in any recognizable form is not a curious artifact of history. It seems to be shared by political conservatives in the United States today. In his unsuccessful campaign for the Republican Presidential nomination in the year 2000, Gary Bauer often remarked that his supporters, political and religious conservatives, looked at the liberal stance in favor of abortion and physician-assisted suicide, particularly the extent to which they not only received serious consideration in the political discourse but had actually become socially acceptable, and as a result could no longer recognize their country. His campaign rhetoric is reminiscent of Feinberg's characterization of the recurrent nightmare of the moral conservative, which is "that he will awaken in a society of barely recognizable aliens, dressed in exotic costumes, speaking a nearly unintelligible tongue, and engaging openly in unsavory though harmless amusements."[14] Of course, the conservatives Bauer purported to represent do not consider abortion, physician-assisted suicide, or tolerance of homosexual relationships to be "harmless amusements." Nevertheless, at least one of their claims is that if there is a "moral majority" view that would not even countenance a serious debate of the right to obtain an abortion or physician-assisted suicide, or to engage in homosexual relations between consenting adults, then a society that does so is a different society, one whose existence could have and should have been prevented by the legal enforcement of the moral majority's views. It is important to notice that the Devlin position is not merely an effort by moralistic busybodies to paternalistically protect careless or improvident people from their own folly. The object of concern is not really the welfare of moral transgressors, but that of the society itself, as embodied by the moral majority.

The other side of this great debate over the legal enforcement of morality has been led by H.L.A. Hart. He has done so by espousing essentially the views of the great Nineteenth Century British philosopher John Stuart Mill as expressed in his seminal essay *On Liberty:* "the only purpose for which power can rightfully be exercised over any member of a civilized community against his will is to prevent harm to others."[15] There are two important dimensions to Mill's proposition. The first is the argument against paternalism, wherein a society intercedes in the conduct of another for her own good. The evil of this approach, according to Mill, is that a free individual must be at liberty to pursue the good as she defines it, unimpeded by conflicting, or more importantly, prevailing societal perceptions of how it is constituted. This is the "tyranny of the majority" which is unacceptable in a free society. The second dimension relates to Devlin's defense of the societal right of self-preservation through the legal enforcement of moral standards prevailing in a society. Mill, as well as Hart, would dispute the contention that actions which merely harm the individual engaging in them, or which contravene societal standards but without demonstrable harm to others, can in a free and open society legitimately be made the subject of criminal or civil penalty.

As we consider at length the discussion of moral issues in major bioethics litigation, we will revisit the question of how harm is defined or determined when individual liberty is challenged by interests of society which partake of moral norms or values. Not surprisingly, the most contentious issues have found their way into Constitutional discourse and decisions of the Supreme Court.

5. LAW AS THE MIDWIFE AT THE BIRTH OF BIOETHICS

The actual or proper relationship between law and morality is of more than mere historical significance to health care. At least in the United States, many health care professionals and bioethicists have expressed dismay and consternation over the extent to which law appears to be driving bioethical thinking. Such concerns are not merely the product of the medical profession's historic disdain for attorneys and their role in medical malpractice litigation.

Others – usually, but not always, those who are themselves trained in the law – see the extent of the law's influence as a natural and quite understandable consequence of certain significant historical events. Arthur Caplan, for example, declares that "bioethics was born from the ashes of the holocaust."[16] Caplan is referring, of course, to the Nuremberg trials, and more particularly the second trial of the Nazi physicians who conducted "medical research" on those imprisoned in the concentration camps. Out of this trial (of multiple defendant physicians) was fashioned by the judges a set of ten principles for medical research on human subjects that has come to be known as "The Nuremberg Code." The hallmark of the Code is its first principle: "The voluntary consent of the human subject is absolutely essential." As the Code's elaboration of this principle makes clear, what is required is not simple consent (acquiescence) that was at the time customary in clinical practice, but rather informed consent based upon a full and accurate disclosure of "the nature, duration, and purpose of the experiment; the method and means by which it is to be conducted; all inconveniences and hazards reasonably to be expected; and the effects upon his health or person which may possibly come from his participation in the experiment."[17]

While the Code has been referred to as a foundational document in bioethics, it was promulgated by a legal tribunal in consultation with certain physicians. It demanded an unprecedented level of disclosure of detail about procedures and risks, which later, as we shall see in Chapter 4, was adapted by courts in the United States and made binding upon the medical profession in clinical practice. Thus, bioethics inherited the Code and the concept of informed consent as a norm of physician conduct. Neither bioethics nor its precursor, medical ethics, can take any credit for fashioning the moral strictures initially imposed on the clinical and research arms of the medical profession. Indeed, so contrary to prevailing professional practices were the demands of the Nuremberg Code that, as we shall see in Chapter 4, decades later flagrant violations continued to come to light.

Based upon the pivotal role of law in fashioning the norms of behavior in clinical research and practice, George Annas maintains that "American law, not philosophy or medicine, is primarily responsible for the agenda, development, and current state of American bioethics."[18] Moreover, he suggests that "American physicians have a long history of equating ethical obligations with legal obligations."[19] If this assertion is correct, and as our analysis will reveal, it has much to commend it in the historical development of medical jurisprudence and bioethics, it highlights a curious fact about medicine's traditional distaste for and distrust of the law and lawyers. For one thesis of this book is that contemporary medical jurisprudence is largely, though certainly not exclusively, a series of instances in which the law has stepped in to fill an ethical vacuum or to redress a pattern and practice of ethical wrongs. In some cases, such as informed consent, the law was invoked by injured patients; but in other instances, such as a number of the cases on refusal of life-sustaining treatment, physicians and the health care institutions in which they practice invoked the jurisdiction of the courts in pursuit of direction or the cloak of legal immunity for treating the patient over her objections. These cases will be the focus of Chapter 8. The question thus presented is why health care institutions and professionals would be asking the courts to tell them what is the appropriate way to resolve disputes between providers and patients on the course of treatment. Should not the ethics of the profession address this issue with sufficient breadth and clarity to cover all but the most idiosyncratic of situations? And if professional ethics does not, perhaps because of novel issues raised by the revolution in medical technology, then cannot the best minds in health care ethics be brought together to fashion new principles or policies?

The cynical answer to such rhetorical questions has already been offered by George Annas. To establish ethical guidelines for the resolution of contentious cases, and then to follow them consistently, requires the courage of one's convictions, i.e., moral courage. According to Annas: "One looks in vain for any history of either moral courage or ethical self-policing in American Medicine. Instead what one discovers is a consistent pattern of special pleading and self-interest."[20] He offers as a case in point the response by the American Medical Association to the increasing public demands that physicians readily stop and give aid to the victims of accidents, which was to lobby the state legislatures for legal immunity from malpractice liability for physicians who responded in this manner. There is something disconcerting, if not unseemly, about members of a profession, the core value of which is beneficence, attempting to extract a legal *quid pro quo* before acknowledging and undertaking a moral duty to those afflicted with illness. Courts have, for example, uniformly rejected efforts by health care institutions and professionals to obtain a waiver of liability from patients as a condition for receiving indigent care.[21] More significant, however, is the proposition implicit in modern medicine's almost mechanical resort to the law when difficult issues of professional ethics present themselves. One interpretation is that there is a consensus within the medical profession that ethical principles, even when codified, as many of them are, in the American Medical Association Code of Medical Ethics, constitute an insufficient basis upon which to base professional conduct. The fear, one suspects, is that if the professional conduct is challenged the law will, more likely than not, decide the case in a manner that is inconsistent with the professional ethic. As we shall see in Chapter 4, that is precisely what happened in the matter of information disclosure.

Both the emergence of bioethics as a discipline and a practice and the transformation of medical jurisprudence from the narrow considerations of forensic medicine to the broader considerations of patient rights were fueled by a sense of moral outrage over particular conduct by physicians or aspects of medical treatment or research.[22] Interestingly, however, the primary candidates for the date of birth of bioethics have not traditionally been, with one exception, as I will suggest they should be, seminal cases in the new medical jurisprudence. While a plausible candidate, as I have earlier intimated, might be the crafting of the Nuremberg Charter, the fact remains that decades passed before anything remotely resembling bioethics began to take place. Passing over the birth of informed consent in 1957, which will be the focus of Chapter 4 and which no one has suggested as a candidate for this birth date, the first viable candidate is the 1962 article in *Life* magazine by Shana Alexander on the innovative Seattle kidney dialysis program. The public furor generated by the article over the role of the so-called "God Squad," a committee which purported to select the few patients who would receive this new therapy based upon their perceived social worth, provided foreshadowing of the potential power of bioethical issues to energize public debate.[23] Interestingly, that same year brought the revelations of the widespread fetal harm caused by the use of the drug Thalidomide by pregnant women. The saga of one particular patient, Sherri Finkbine, and her pursuit of a therapeutic abortion in order to avoid giving birth to a child severely deformed by the drug, was chronicled in the press.[24] Again, part of the public outrage pertained to the conduct of a hospital committee in passing upon the propriety of her decision to abort the fetus and her ultimate need to travel to a Scandinavian country in order to have the procedure performed. We will consider that case further in Chapter 7.

Over ten years passed before the next set of viable candidates for the birth of bioethics appeared: the exposure of the Tuskegee Syphilis Study in 1974 (which we will consider in Chapter 4), the impetus to create the National Commission for the Protection of Human Subjects of Biomedical and Behavioral Research in 1975, and the litigation involving Karen Ann Quinlan in 1976. Again, it is the level of public furor over these events, and the felt need for a means of assessing and addressing the "wrongness" of behaviors which had been blessed by influential members of the medical profession, which renders them plausible candidates for the birth of bioethics.

For the most part, this pursuit of a consensus on the time when bioethics was born is a futile one. Bioethics gradually came into being as a process, not as the result of a single event or even a cluster of events. One of the recurring themes of this book is the singular role of the law in creating a space for bioethical discourse about medical research and patient care. For centuries, medicine in all of its forms was a closed system. Only physicians could judge the reasonableness, appropriateness, and morality of what other physicians or health care institutions did. Beginning with the Nuremberg doctors trial, and continuing with judicial decisions about informed consent and the right to refuse treatment, the courts asserted and defended the proposition that no profession or social institution can be left entirely free to police itself. Judges had the temerity to insist that physicians submit themselves to the scrutiny of non-physicians in matters of major public importance. The law created the moral and political space into which bioethics could enter and lay claim to a legitimate role in the analysis of medical issues with moral implications, as well as in indicating when those implications were present.

My own candidate for the birth of a genuinely holistic medical jurisprudence, by which I mean one that moves far beyond the narrow considerations of forensic medicine and professional negligence described in Chapter 1, is the judicial recognition and imposition of the doctrine of informed consent. But here too, as we shall find, it is both misleading and inaccurate to focus on a single case, for the seeds of the concept of informed consent were sown decades before the term was actually used.

ENDNOTES AND REFERENCES

1. See, e.g., Brian Bix, "Natural Law Theory," in Dennis Patterson, ed., *A Companion to Philosophy of Law and Legal Theory* (Oxford: Blackwell Publishers, Inc., 1999), pp. 223-240.
2. In his critique of H.L.A. Hart's positivistic viewpoint, Dworkin argues that what that viewpoint ignores is "the crucial fact that jurisprudential issues are at their core issues of moral principle, not legal fact or strategy." Ronald Dworkin, *Taking Rights Seriously* (Cambridge: Harvard University Press, 1977), p. 7.
3. Ronald Dworkin, *Freedom's Law* (Cambridge: Harvard University Press, 1996), p. 12.
4. Two examples of this kind of argument are offered by Roger B. Dworkin and Richard A. Posner. The former actually maintains that "there was no need for the Supreme Court to decide anything about abortion." Roger B. Dworkin, *Limits – The Role of Law in Bioethical Decision Making* (Bloomington: Indiana University Press, 1996), p. 53. The latter argues that in deciding *Roe v. Wade* the Supreme Court "ducked the moral issue." Richard A. Posner, *The Problematics of Moral and Legal Theory* (Cambridge: Harvard University Press, 1998), p. 134.
5. John Austin, *The Providence of Jurisprudence Determined* (London: Weidenfeld & Nicholson, 1955).
6. H.L.A. Hart, *The Concept of Law* (Oxford: The Clarendon Press, 1961), pp. 49-64.
7. Oliver Wendell Holmes, "The Path of the Law," *Harvard Law Review* (1897), 10: 61-80.
8. Holmes, note 7, p. 63.
9. *Trial of the Major War Criminals before the International Military Tribunal*, Vol. 1, 11, 22 (Nuremberg, 1947-1949).
10. *Trial*, note 9, Vol. 19, pp. 465-466.
11. H.L.A. Hart, "Positivism and the Separation of Law and Morals," *Harvard Law Review* (1958), 71: 593-629.
12. Lon Fuller, "Positivism and Fidelity to Law," *Harvard Law Review* (1958), 71: 630-672.
13. Patrick Devlin, *The Enforcement of Morals* (Oxford: Oxford University Press, 1959).
14. Joel Feinberg, *Harmless Wrongdoing* (Oxford: Oxford University Press, 1990), p. 74.
15. John Stuart Mill, *On Liberty* (New York: The Liberal Arts Press, Inc., 1956), p. 13.

16. Arthur Caplan, Keynote Address, Holocaust and Bioethics Conference, Minneapolis, MN. (1989), cited in George J. Annas, "The Dominance of American Law (and Market Values) Over American Bioethics," in Michael Grodin, ed., *Meta Medical Ethics: The Philosophical Foundations of Bioethics* (Dordrecht: Kluwer Academic Publishers, 1995), p. 85.
17. Germany (Territory Under Allied Occupation, 1945-1955: U.S. Zone) Military Tribunals, 1947, "Permissible Military Experiments." In Vol. 2 of Trials of War Criminals Before Nuremberg Tribunals Under Control Law, No. 10, pp. 181-184, Washington, D.C.: U.S. Government Printing Office.
18. George J. Annas, *Standard of Care: The Law of American Bioethics* (New York: Oxford University Press, 1993), p. 3.
19. Annas, note 18, at p. 86.
20. Annas, note 18, at p. 87.
21. Tunkl v. Regents of the University of California, 383 P.2d 441 (Cal. 1983); Emory University v. Porubiansky, 282 S.E. 2d 903 (Ga. 1981); Ash v. New York University Dental Center, 564 N.Y.S.2d 308 (N.Y. App. Div. 1990).
22. Charles E. Rosenberg, "Meanings, Policies, and Medicine: On the Bioethical Enterprise and History," *Daedalus* (1999), 128: 27-46, p. 37.
23. Shana Alexander, "They Decide Who Lives, Who Dies: Medical Miracle Puts a Burden on a Small Committee," *Life* 53 (102): November 9, 1962.
24. For a detailed chronology of the Finkbine case and its coverage in the media *see* David J. Garrow, *Liberty and Sexuality* (N.Y.: MacMillan Publishing Co., 1994), pp. 285-289.

Chapter 4

LAW AND THE PHYSICIAN-PATIENT RELATIONSHIP: INFORMED CONSENT IN THEORY AND PRACTICE

1. FROM ANCIENT TO EARLY MODERN MEDICAL ETHICS

The figure of Hippocrates of Cos (460-370 B.C.) looms large in the history of Western medicine in general and in the conceptualization of a physician's responsibility to the patient in particular. What is singularly problematic about the influence of Hippocrates upon medicine is that it has been based as much, if not more, on myth and misinformation as upon fact. The most important case-in-point is the so-called "Hippocratic Oath." Recent scholarship has persuasively demonstrated that many of the most often-cited ethical admonitions set forth in the Oath – against the giving of a deadly drug if asked, or suggesting it, against providing a woman with an abortive remedy, against performing surgery – are inconsistent with the practices of Hippocrates and most other physicians during the Hippocratic era and after. The best explanation for this anomaly has been offered by the historian Ludwig Edelstein, who opined that the ethical pronouncements of the Oath strongly suggest that it is essentially a manifesto by a Pythagorean cult, written at least a generation after the Hippocratic era.[1] So compelling has been the purported Hippocratic origin of the prohibitions against abortion and physician-assisted dying, that in writing the opinion of the Court in *Roe v. Wade*, Justice Blackman felt the need to cite Edelman's research undermining the authenticity of the Oath.[2]

What is most interesting in terms of the subject matter of this chapter is that while the Oath contains an admonition that the physician maintain the confidentiality of that which is learned in the context of the physician-patient relationship, it contains no admonition that the physician be honest and forthright in disclosing important information to the patient. In this regard at least, the Oath is consistent not only with the Hippocratic corpus (of which 70 treatises of diverse length remain), but with every other significant treatise of medicine or code of medical ethics written subsequently until quite late in the Twentieth Century.[3] The inescapable conclusion is that a candid and comprehensive disclosure by the physician to the patient regarding diagnosis, prognosis, options for treatment, likelihood of success, and reasonable alternatives, was not considered to be a part of the physician's professional or ethical responsibility. To the extent that disclosure of relevant information by physician to patient was discussed at all in the medical literature, its thrust was toward nondisclosure. Indeed, the customary withholding of information from patients was justified by physicians on the principle of nonmaleficence, since it was presumed that the disclosure of detailed information about the patient's illness, the risks of therapy, and the possibility or even probability that the treatment might fail, would be likely to harm the patient. We will explore this point further in a subsequent section on the therapeutic privilege.

2. THE PHYSICIAN AS FIDUCIARY

Viewing the physician-patient relationship from a legal rather than a traditional medical perspective, medicine's legacy of silence is difficult to fathom or to justify. Like the attorney-client and other relationships of trust between professionals and those in society to whom they provide a service, the physician-patient relationship is fundamentally fiduciary in nature. A fiduciary is "one who owes to another the duties of good faith, trust, confidence, and candor" and "one who must exercise a high standard of care" in discharging their duties.[4] A classic instance of a fiduciary is someone who undertakes to manage the financial assets or property of another. It is virtually inconceivable that a reasonable person would confer such an important responsibility upon another with the expectation that the fiduciary would adopt a pattern and practice of nondisclosure of their strategy for managing the assets or the details of transactions involving those assets. Candor is the hallmark of such a relationship because one cannot have trust and confidence that another is acting responsibly on one's behalf without regular, candid disclosure of pertinent information. It would be extremely difficult for a fiduciary to plausibly argue that it was not in the fiduciary's best interests to be well-informed on such matters, or at least as fully informed as they wished to be. Yet that is precisely the philosophy of medical practice throughout most of its history.

The paternalistic paradigm in the traditional physician-patient relationship casts the physician in the role of father and the patient in the role of child in the quintessential "Father Knows Best" fashion. The basic presupposition is that the physician (like a good father) has the patient's (child's) best interests at heart, and that the patient is as ill-equipped as a child to understand and appreciate what those interests might be. This unique medical slant on the capacity of the typical patient to determine his own best interests suggests why the medical profession has had such a different perspective on the physician-patient relationship than the legal profession has had on the attorney-client relationship. While attorneys do not presume that their clients know any more about the law than patients know about medicine, they do presume that clients are adults with the right to make decisions about their personal affairs. This is the key presumption about the fiduciary relationship that appears to have been missing in the traditional physician-patient relationship.

3. EARLY MODERN MEDICAL JURISPRUDENCE AND CONSENT

As others who have surveyed the legal history of medical jurisprudence have concluded, Nineteenth Century case law presents very few cases involving issues of consent, and most of those actually involved negligent treatment, which turned out to be the primary basis upon which professional liability was found.[5] Early in the Twentieth Century, a number of decisions appeared with language that would prove to be critical to the ultimate formulation of the doctrine of informed consent. In the first of these cases, *Mohr v. Williams*,[6] the patient Anna Mohr gave her consent to a surgical procedure on her right ear. However, in the midst of the operation the surgeon determined that it was the left ear which posed the problem, and he proceeded to perform the procedure on it. When it became clear to the patient that the operation had actually worsened the condition, she filed a battery action (a tort claim based upon wrongful or nonconsensual touching) alleging lack of consent to the procedure actually performed. The precedential significance of the case lies in the following language from the opinion of the court:

> The free citizen's first and greatest right, which underlies all others – *the right to himself* – is the subject of universal acquiescence, and this right necessarily forbids a physician or surgeon . . . to violate without permission the bodily integrity of his patient by a major or capital operation . . . without his consent or knowledge.
>
> * * *
>
> If a physician advises a patient to submit to a particular operation, and the patient weighs the dangers and risks incident to its performance, and finally consents, the patient thereby, in effect, enters into a contract

> authorizing the physician to operate *to the extent of the consent given, but no further.*[7]

Although it appears in the opinion without attribution, the phrase "the right to himself" would appear to be a direct reference to the often-quoted assertion by the philosopher John Locke that "every man has a perfect freedom to order their acctions and dispose of their persons as they think fit."[8] The same philosophical proposition has been much more recently articulated by law professor and jurist Charles Fried in the form of "every person belongs to himself, not to anyone else and not to society."[9] The foundational ethical principle (discussed in Chapter Two) underlying these pronouncements is that of respect for individual autonomy.

What is particularly interesting about the opinion is that while the facts of the case clearly indicate that there was no consent whatsoever to the operation on the patient's left ear, the court nonetheless introduces language relating to and presupposing the disclosure of information by the physician to the patient about the risks of the procedure. Thus, while the case was one of nonconsent rather than lack of informed consent, this court intimates that it would be receptive to such a claim if it were supported by the facts in evidence.

Another early Twentieth Century case is *Pratt v. Davis*[10] in which a patient sued a surgeon for performing a hysterectomy without her consent after she had engaged him to treat her epilepsy. The surgeon defended his conduct by arguing that if a patient engages a surgeon to provide treatment, then there is an implied consent to provide whatever treatment he deems necessary. Interestingly, the court soundly rejected this attempt to greatly expand the implied consent doctrine.[11] However, a detailed consideration of the facts of the case reveals much about the mind-set of the medical community in this pre-informed consent era. The surgeon unabashedly argued that he had acted properly when he told the patient only that he was going to repair a few cervical and rectal tears when in fact he intended to remove her uterus and ovaries in an effort to cure her epilepsy. This deception was necessary because he wanted her "to come to the operating room without violence." While the physician's patronizing and paternalistic attitude toward the patient appears shocking now, and caused even the court reviewing the case in 1905 some degree of consternation, there is nothing about his conduct that would have violated the professional norms of the medical profession at that time.

The most celebrated of the early cases, perhaps because of the eminent status of the judge who wrote the opinion (Benjamin Cardozo), is *Schloendorff v. The Society of New York Hospital*.[12] Unlike *Mohr* and *Pratt*, *Schloendorff* was a suit brought by a patient against a hospital rather than the physician. The patient sought to hold the hospital liable for a surgical procedure performed without her consent. Although she had consented to an examination under anesthesia, she had insisted that no surgical procedure be performed. Despite her clear instructions, once she was under the effects of the ether the surgeon proceeded to remove a fibroid tumor. Because the court declined to hold the hospital directly or vicariously liable for the physician's conduct (something which many courts have been willing to do in the last half of the Twentieth Century), the remarkable language for which the decision has become famous was mere *dictum* (an expression of opinion not essential to the holding of the case before the court). That language reads as follows: "Every human being of adult years and sound mind has a right to determine what shall be done with his own body; and a surgeon who performs an operation without his patient's consent commits an assault, for which he is liable in damages."[13]

What has been characterized as "a classic statement of the patient's right to self-determination,"[14] is, I suggest, a grand and lofty pronouncement that has, in the decades since it was written, died the death of a thousand qualifications. About this I will have more to say later. For the present, the significance of these early decisions is that courts were willing to characterize as flagrant violations of the fundamental rights of patients conduct on the part of physicians that would not have subjected them to disciplinary action by medical boards for unprofessional or unethical conduct. This is not so much because the boards lacked the authority, but because such conduct reflected the custom and practice of physicians. It would not have occurred to the typical physician board member that any unprofessional conduct had taken place. We can begin to discern the significant disconnect between the ethics of the medical profession in its traditional stance and the expectations of the larger society. The gap began to close only after the jurisdiction of the courts was invoked more frequently, in the decades to follow, in the name of patients' rights.

4. FROM CONSENT TO INFORMED CONSENT

As we noted in the previous section, while some of the early court opinions employ language suggesting that a patient's consent to treatment should be based upon the disclosure of relevant information, an analysis of the facts of these cases indicates that each was an instance in which there was no consent at all, informed or otherwise, to the treatment that was administered. Over fifty years were to pass before a reported case would address the issue of a physician's responsibility to disclose any detailed information about the procedure for which the patient's consent is being sought.

Martin Salgo received a preliminary diagnosis of occlusion of the abdominal aorta after being referred to a faculty physician at Stanford University for symptoms of pain and weakness in his legs. A translumbar aortography (the injection of the contrast material sodium urokon into the aorta to locate blockage) was determined to be necessary to confirm the diagnosis, and it was subsequently performed under general anesthesia. On the morning following the aortography, Mr. Salgo reported that his lower extremities were paralyzed, a condition which turned out to be permanent.

A malpractice suit was filed against the university, the physician who ordered the aortography, and other medical personnel involved in the performance of the procedure, resulting in a judgment in favor of the plaintiff. Among the jury instructions was the explication of the duty of a physician to disclose to the patient "all of the facts which mutually affect his rights and interests and of the surgical risk, hazard, and danger, if any . . ."[15] An issue raised on appeal of the judgment by the defendants was the extent of the disclosures made to Mr. Salgo about the aortography procedure. Salgo, his wife and son testified that he was never advised that one would be performed. The defendant physicians asserted that the aortography procedure was recommended to the patient and that he consented to it. However, they also conceded that risks of the procedure were not discussed with him.

In crafting its own formulation of the duty to disclose, the appellate court was clearly seeking to balance the patient's need for pertinent information upon which to base a decision with the need to confer a reasonable amount of discretion upon the physician to tailor the disclosure to the vagaries of particular patients. Consequently, the court arrived at the following formulation: "A physician violates his duty to his patient and subjects himself to liability if he withholds any facts which are necessary to form the basis of an intelligent consent by the patient to the proposed treatment." Later in the opinion, as the court discussed the appropriate exercise of discretion by the physician in framing the disclosure dialogue, "intelligent consent" became "informed consent."[16] Implicit in the holding of the court is the basic proposition that an uninformed consent is really no consent at all. The very concept of consent presupposes a minimally sufficient understanding of that to which consent is being given. Also intrinsic in the concept of informed consent, one would suppose, is the converse – informed refusal. However, as we shall see in Chapter 8, a great many cases had to be litigated before the patient's right to refuse life-sustaining medical treatment was recognized as a fundamental principle of medical jurisprudence.

The transition from the recognition of a physician's duty to obtain a patient's consent to the recognition of a physician's duty to obtain a patient's *informed* consent also brought about a change in the nature of the theory of liability from battery to negligence. What would have to await subsequent decisions would be a clear and definitive delineation of the types of information that are essential to a truly informed consent by a patient. With such a delineation would come a determination of the extent to which expert medical testimony would be necessary in order to establish that the duty to disclose had or had not been fulfilled. Traditionally, a physician's professional duties are determined by the custom and practice of physicians of similar training and experience when caring for similarly situated patients. One of the formidable obstacles that Mr. Salgo would have had to overcome at trial would have been a contention on the part of the defendants, and presumably their expert witnesses, that the custom and practice of physicians in 1957 was not to disclose the risks of treatment to patients, based upon the time-honored concern of physicians that such disclosures only serve to needlessly frighten patients. The presence of the above-quoted jury instruction on the duty to disclose indicates that somehow the plaintiff managed to persuade the court, perhaps through reliance on precedents such as *Pratt, Mohr,* and *Schloendorff*, that the duty to disclose existed in spite of the medical profession's failure to honor it in clinical practice. We shall see in the analysis of the next major informed consent decision that the plaintiff was not so fortunate at the level of the trial court, but was ultimately vindicated on appeal.

In *Natanson v. Kline*[17] the plaintiff, Irma Natanson, alleged that the treatment that she had received for breast cancer (cobalt irradiation) was negligent in a number of different ways, including the extent to which she had been warned of its risks. As a result of this negligent treatment, she sustained serious injury. After having presented evidence at trial that the defendant, Dr. Kline, had failed to warn his patient that her treatment involved significant risk of bodily injury and even death, the plaintiff requested, but the trial judge refused to give, the following jury instruction:

> You are instructed that the relationship between physician and patient is a fiduciary one. The relationship requires the physician to make a full disclosure to the patient of all matters within his knowledge affecting the interests of the patient. Included within the matters which the physician must advise the patient are the nature of the proposed treatment and any hazards . . . which are known to the physician. Every adult person has the right to determine for himself or herself whether or not he will subject his body to the hazards of any particular medical treatment.
>
> You are instructed that if you find from the evidence that the defendant Kline knew that the treatment he proposed to administer to plaintiff involved hazard or danger he was under a duty to advise plaintiff of that fact and if you further find that the defendant Kline did not advise plaintiff of such hazards then defendant Kline was guilty of negligence.[18]

The last sentence of the first paragraph of the proposed instruction has clearly been extracted almost verbatim from Cardozo's opinion in *Schloendorff*. The second paragraph presents nothing more than a duty to disclose *known* risks, as opposed to actual risks, whether or not the physician is aware of them. The trial court ruled in favor of the defendants on all counts of the complaint and the plaintiff appealed.

The appellate court deciding *Natanson* makes the following interesting observation: "We are here concerned with a case where the physician is charged with treating the patient without consent on the ground the patient was not fully informed of the nature of the treatment or its consequences, and, therefore, any 'consent' obtained was ineffective."[19] This is a clear articulation by the court of the fundamental proposition inherent in the doctrine of informed consent, i.e., that an uninformed consent is no consent at all. The basis of this assertion is that an uninformed consent is not an authentic expression of the patient's individual autonomy. However, the court does not proceed to the next logical step, which would be to treat the case as one of battery. Instead, because Mrs. Natanson consented to the treatment, albeit without sufficient information, the case is treated as a negligence claim. As we shall see, the implications for the plaintiff with regard to the burden of proof and the need for expert testimony are significant.

The most frequently-quoted language from *Natanson* is the following:

> Anglo-American law starts with the premise of thorough-going self determination. It follows that each man is considered to be master of his own body, and he may, if he be of sound mind, expressly prohibit the performance of life-saving surgery, or other medical treatment. A doctor might well believe that an operation or form of treatment is desirable or necessary but the law does not permit him to substitute his own judgment for that of the patient by any form of artifice or deception.[20]

This is a sweeping statement. It elevates respect for individual autonomy to a first principle of American law. Furthermore, it recognizes the right of an adult with decisional capacity to refuse all medical treatment, including that necessary to save one's life. Finally, it precludes a physician, even with the most beneficent of motives, from manipulating the disclosure of information so as to prevent the patient from making what is in the physician's judgment a "wrong" decision.

The *Natanson* court provides more detail than did the court in *Salgo* about the physician's duty of disclosure:

> The duty of the physician to disclose . . . is limited to those disclosures which a reasonable medical practitioner would make under the same or similar circumstances. How the physician may best discharge his obligation to the patient in this difficult situation involves primarily a question of medical judgment.
>
> In considering the obligation of a physician to disclose and explain to the patient in language as simple as necessary the nature of the ailment, the nature of the proposed treatment, the probability of success or of alternatives, and perhaps the risks of unfortunate results . . . we do not think the administration of such an obligation . . . presents any insurmountable obstacles.[21]

Having ventured some of the essential elements of an informed consent, the court then leaves to the discretion of the physician the manner in which the information is to be imparted by noting that the obligation extends no further than "those disclosures which a reasonable medical practitioner would make under the same or similar circumstances."[22] There is a serious problem inherent in this "reasonable physician" standard of disclosure which will become clear in the discussion of the last of the major informed consent decisions that we will consider.

Canterbury v. Spence, a 1972 federal court decision, involved the failure of a surgeon to advise his patient of the risk of paralysis posed by the laminectomy procedure recommended as treatment for severe, persistent back pain. There are many complicating factual matters in *Canterbury,* relating to the actual cause of the patient's permanent paralysis, a discussion of which would carry us far beyond the scope of this analysis. For our purposes, the most important language in the decision of the court is the following:

> There are, in our view, formidable obstacles to acceptance of the notion that the physician's obligation to disclose is either germinated or limited by medical practice. To begin with, the reality of any discernable custom reflecting a professional consensus on communication of option and risk information to patients is open to serious doubt. We sense the danger that what is in fact no custom at all may be taken as an affirmative custom to maintain silence . . . Respect for the patient's right of self-determination on particular therapy demands a standard set by law for physicians rather than one which physicians may or may not impose upon themselves.[23]

What the *Natanson* court completely failed to acknowledge, Judge Robinson in *Canterbury* makes abundantly clear, i.e., the problem inherent in the "reasonable physician" standard of disclosure is that prior to the decision in *Salgo*, physicians were scrupulously maintaining their traditional stance of silence, or at least of not venturing beyond the minimum disclosure necessary to obtain the patient's consent. Their practice was not based upon evil intentions, but it was how they had been trained and enculturated, and it was consistent with the prevailing ethic of the physician-patient relationship. In the words of Jay Katz:

> Physicians have always maintained that patients are only in need of caring custody. In order to accomplish that objective they were obliged to attend to their patient's physical and emotional needs and to do so on their own authority, without consulting with their patients about the decisions that needed to be made. Indeed, doctors intuitively believed that such consultations were inimical to good patient care.[24]

Judge Robinson's insistence upon a standard for disclosure "set by law for physicians" is not, therefore, mere judicial arrogance, but a recognition that a duty to disclose to patients material aspects of diagnosis and treatment had never been voluntarily assumed by physicians and would not be enthusiastically embraced.

Two alternatives to the reasonable physician standard of disclosure presented themselves to the court. One is a subjective standard, i.e., what would this particular patient wish to know? The other is an objective standard, i.e., what would a reasonable patient in this particular patient's situation wish to know? Judge Robinson rejected the subjective standard for two reasons. First, because adoption of the subjective standard would tie liability to an after-the-fact determination by the patient that if the risk which materialized had been warned of in advance, the procedure or therapy would have been declined. The reliability of the patient's testimony in this regard would indeed be troublesome, although not necessarily prohibitively so, inasmuch as a jury is quite capable of weighing the patient's credibility and motivation to give self-serving testimony. The second, and less defensible, reason was that it "would summon the physician to second-guess the patient, whose ideas on materiality could hardly be known to the physician" and "make an undue demand upon medical practitioners."[25] The court fails to explain how it could possibly be more difficult for a physician to inquire of an actual patient, present in the examining room, what his or her informational needs are than to intuit what the informational needs of a hypothetical, reasonable patient might be. It also fails to explain how "the boundaries of the physician's duty to reveal" information can be "measured by the patient's need" under the objective standard articulated. Ironically, one commentator recently noted that "doctors have not unreasonably complained that the vague 'reasonable patient' standard of tort law tells them deplorably little about their duties of informed consent."[26]

According to the standard set forth in *Canterbury*, if the hypothetical reasonable patient would have consented to the procedure when apprised of the risk which actually materialized in the plaintiff, then there would be no liability on the part of the physician despite the fact that the plaintiff would not have consented because in the exercise of her own values she deemed the risk excessive. It is for this reason that while *Canterbury* advanced the cause of patient self-determination beyond *Salgo* and *Natanson*, and far beyond what it had been before the recognition of the informed consent doctrine, it nonetheless constitutes one of the thousand qualifications to Cardozo's assertion of "the right to determine what shall be done with [one's] own body."

5. THE THERAPEUTIC PRIVILEGE

We noted in Section 1 the ostensibly beneficent rationale for the withholding of information from patients – that disclosure would unnecessarily distress them. While the recognition of the informed consent doctrine eliminated this approach as a general rule, the *Canterbury* case and subsequent decisions recognized an exception to the general rule of disclosure called the "therapeutic privilege." The privilege may properly be invoked when a physician has a good faith basis to believe that disclosure of certain information to the patient poses a significant risk of actual harm – physical and/or psychological. In recognizing such a privilege, the *Canterbury* court acknowledged the quite real risk that the exception might expand so as to obliterate the duty to disclose, and cautioned physicians that invocations of the privilege would be carefully scrutinized.

The magnitude of the risk posed by the therapeutic privilege may be much greater than originally anticipated, based upon recent, extensive research by Nicholas Christakis. Christakis has identified an ethos against communicating prognosis in contemporary medical culture.[27] Precisely such an ethos provided the basis for the practice, believed to have since been discredited (at least in the United States), of withholding terminal diagnoses from patients. The new data suggest that the exercise of the therapeutic privilege may actually be more pervasive than imagined.

6. THE PARADIGM SHIFT

The true significance of the line of legal decisions recognizing and developing the doctrine of informed consent is the transformative influence on the model of the physician-patient relationship. Even at the mid-point of the

Twentieth Century, the paradigm for physician-patient interaction was the paternalism that had reigned for over 2000 years. The concept of individual patient autonomy, and its recognition through the exercise of informed consent to and informed refusal of treatment, was virtually unintelligible under this model. Thus, when the duty to disclose to patients pertinent information about the risks, benefits, and alternatives to proposed treatment was imposed upon the medical profession by the courts, the shift to a paradigm that made the concept and the process intelligible was necessary. The new paradigm has become known as "shared decision-making." The physician-patient relationship is analogized to a joint venture in which both parties work cooperatively in pursuit of a common goal. If a mutual understanding cannot be arrived at and maintained, then the alternative is not for one party to assert their greater authority in an effort to coerce the other into compliance, but rather for the joint venture to dissolve because an essential element of its existence is absent.

During the concluding decades of the Twentieth Century medical ethics, as codified in the American Medical Association Code of Medical Ethics (hereinafter AMA/CME), came to formally recognize the paradigm shift brought about by the line of informed consent decisions. Interestingly, however, the 150th anniversary edition of the AMA/CME presents the "fundamental elements of the patient-physician relationship" (note the transition from physician-patient to patient-physician) as though they had an ancient and unbroken lineage in medical practice:

> From ancient times, physicians have recognized that the health and well-being of patients depends upon a collaborative effort between physician and patient. Patients share with physicians the responsibility for their own health care. The patient-physician relationship is of greatest benefit to patients when they bring medical problems to the attention of their physicians in a timely fashion, provide information about their medical condition to the best of their ability, and work with their physicians in a mutually respectful alliance.[28]

The above language to the contrary notwithstanding, the philosophy underlying the AMA/CME's recognition of patient rights "to receive information from physicians and to discuss the benefits, risks, and costs of appropriate treatment alternatives . . . to make decisions regarding the health care that is recommended by his or her physician . . . [and] to courtesy, respect, dignity, responsiveness, and timely attention to his or her needs" is in diametric opposition to the early editions of the AMA/CME which counseled physicians to grant "reasonable indulgence . . . to the mental imbecility and caprices of the sick," and admonished patients that "obedience to the prescriptions of his physician should be prompt and implicit" and that "he should never permit his own crude opinions as to their fitness to influence his attention to them."[29] Only since 1981 has the AMA/CME recognized that "the patient's right of self-decision can be effectively exercised only if the patient possesses enough information to enable an intelligent choice" and that "the patient should make his or her own determinations on treatment." Curiously, the physician's duty to obtain a patient's informed consent to treatment is characterized by the AMA/CME not as an obligation grounded in principles of medical ethics, but as a "social policy." Thus, the AMA/CME maintains, it is not prevailing views of medical ethics, but rather "social policy [which] does not accept the paternalistic view that the physician may remain silent because divulgence might prompt the patient to forgo needed therapy." What is even more intriguing, in light of the insistence of the courts and the medical profession upon the need for an objective standard of disclosure, whether it be that of the reasonable physician or the reasonable patient, is that the AMA/CME nonetheless acknowledges that "[r]ational, informed patients should not be expected to act uniformly, even under similar circumstances, in agreeing to or refusing treatment."[30]

The continuing characterization of informed consent as a matter of social policy by the largest and most influential professional medical organization in the United States places squarely before us the larger question regarding the ultimate significance of whether informed consent and the respect for patient autonomy which it is intended to foster is deemed a matter of law, ethics, or social/public policy. One of the explanations for why patient autonomy, and hence informed consent, became a significant legal issue in the U.S. in the 1960's and 1970's was that it developed as a natural outgrowth of the civil rights, consumer rights, and women's rights movements. "Rights talk" and demands for the recognition and enforcement of newly recognized rights tends to lead naturally (at least in the U.S.) to legislation, litigation, or both. Thus, social and public policy both influences and is influenced by the legal process. But prior to the proliferation of litigation concerning patient rights, medical ethics had generally been insulated from and relatively impervious to these policy influences. Indeed, what characterizes the last half of the Twentieth Century and the rise of bioethics is the unprecedented influence that social and public policy and law have exerted over both clinical research and practice.[31]

In their highly-regarded text on the development of informed consent, Faden, Beauchamp and King assert: "Notwithstanding the societal decision to resolve certain limited kinds of problems in the courts, informed consent in clinical practice should be first a problem of medical ethics, not legal requirements."[32] "Should" is the operative word. The authors express a desideratum, not a fact. As they go on to acknowledge, the law was endowed with internal principles conducive to the concept of informed consent, e.g., self-determination, bodily integrity and fiduciary responsibilities, which medicine lacked. Consequently, as they acknowledge: "by comparison to the contributions made by physicians and indeed writers in any field, including ethics, the courts must be said to have been the pioneers of today's concerns about informed consent."[33] There is another reason why this has been the case, which will be the focus of Chapter 10, and to which I will only briefly allude at this point. Codes of Ethics such as that of the AMA are admonitory only. They purport to represent and articulate a consensus of the profession with regard to the parameters of ethical behavior. Noncompliance, whether intentional or inadvertent, carries no demonstrable penalties. Unless a state medical licensing board were to take disciplinary action against physicians when their professional conduct violates the AMA/CME, even wholesale disregard for those principles would have no direct, immediate, and adverse consequences for the physician. Such is not the case when the law recognizes a principle. There are consequences which can and will be made to follow directly from its disregard. Most Americans recognize this fact, and they proceed accordingly.

A vivid case-in-point recently arose in the State of California. The daughter of an elderly patient who experienced a painful death from lung cancer filed a complaint with the state medical board charging the physician responsible for her father's care with unprofessional conduct for his failure to properly manage his patient's pain in the last days of life. California's medical board has guidelines admonishing physicians to make pain relief a priority in all patients, and the legislature enacted a statute known as the "Pain Patient's Bill of Rights."[34] Following its investigation of the complaint, the board concluded that the subject physician's management of the patient's pain had been inadequate, but it declined to take any disciplinary action whatsoever against the physician. As a result of the board's refusal to act, the complainant retained legal counsel and filed a medical malpractice action against the treating physician.[35]

7. NUREMBERG'S INFORMED CONSENT LEGACY

While it is technically correct to credit the court in *Salgo* with coining the term "informed consent," as we have previously noted, the concept of informed consent was articulated ten years earlier by the Nuremberg tribunal, in the context of the trial of the Nazi doctors. The first principle of the Nuremberg Code provides: "The voluntary consent of the human subject is absolutely essential." That principle is then elaborated at length in the following particulars:

> This means that the person involved should have legal capacity to give consent; should be situated so as to be able to exercise free power of choice, without the intervention of any element of force, fraud, deceit, duress, overreaching, or other ulterior form of constraint or coercion;

> and should have sufficient knowledge and comprehension of the elements of the subject matter involved as to enable him to make an understanding and enlightened decision. This latter element requires that before the acceptance of an affirmative decision by the experimental subject there should be made known to him the nature, duration, and purpose of the experiment; the method and means by which it is to be conducted; all inconveniences and hazards reasonably to be expected; and the effects upon his health or person which may possibly come from his participation in the experiment.[36]

Clearly, an uninformed or an involuntary consent would not meet the requirements of the Nuremberg Code. The other nine provisions of the Code pertain to the design and conduct of the research protocol, the degree of acceptable risk, the qualifications of the investigators, and the duty to discontinue the project when there is evidence of actual or significant potential harm.

As with therapeutic interventions, the need for informed consent to medical research was not recognized and recommended by most physicians. In part, this was the result of a perception that the atrocities perpetrated on the innocent by the Nazi doctors in the name of medical research were so extreme that they would never be at risk of happening again. The reality however, which did not come into full public view for almost 20 years, was that violations of the principles of the Code were widespread and persistent in medical research conducted under the auspices of highly respected scientists and institutions.[37] Among those which ultimately came to light and incited calls for new federal regulations were the artificial induction of hepatitis in an institution for mentally defective children (the Willowbrook State School) under the direction of an investigator who would subsequently become the Chair of Pediatrics at New York University and a recipient of the Lasker Prize, the most prestigious award for research in the United States. A second was the injection of live cancer cells into 22 subjects who were not informed of the nature of the cells as part of a study of immunity to cancer by an investigator who became the President of the American Association for Cancer Research.[38]

In the face of this graphic evidence that the "best and the brightest" medical researchers did not conduct themselves as though the Nuremberg Code had any relevance to their investigations whatsoever, Congress was finally able to overcome the vociferous and sustained opposition of the medical research community and create the National Commission for the Protection of Human Subjects of Biomedical and Behavioral Research in 1974. What helped galvanize the determination of Congress was the public outrage over a 42-year research project sponsored by the U.S. Public Health Service, "The Tuskegee Study," which carefully followed the progression of untreated syphilis in a group of African American men in Macon County, Alabama. The research subjects were not told that they suffered from a condition for which treatment had become available long before the study was discontinued, or that the spinal taps which they received were not treatment but merely a means of monitoring their untreated condition.[39]

Ultimately, Congress charged the Department of Health and Human Services with the legal responsibility to promulgate and enforce detailed regulations on the conduct of research on human subjects that are binding on all federally- funded protocols. Integral to the implementation and enforcement of the HHS regulations which were ultimately adopted is the local (i.e., within a particular research institution or consortium of institutions) review of research protocols involving human subjects by an Institutional Review Board (IRB).[40] The federal regulations require the IRB to evaluate proposed research through the application of eight criteria. In order to approve a protocol, the IRB must ascertain that risks to human subjects have been minimized and are commensurate with the benefits to them, if any, as well as to the anticipated significance of the scientific knowledge to be acquired. Reasonable precautions for subject safety, privacy, and confidentiality must be taken, especially for vulnerable subjects, and informed consent must be obtained and documented.[41] The requirements for what must be disclosed in order to ensure an adequately informed consent to participate as a research subject constitute the most detailed provisions of the federal regulations.[42]

Nevertheless, a growing literature suggests a number of significant barriers to the achievement of truly informed consent by research subjects, particularly those who are candidates for Phase I drug trials, most of whom have a terminal condition. These barriers have much more to do with the peculiar vulnerabilities of patients with advanced diseases such as cancer and AIDS than they do with a general disdain for the informed consent process on the part of the medical research community. It has become increasingly clear, for example, that although subjects in Phase I cancer drug trials have been informed they should expect to receive no medical benefit from their participation, they continue to indicate that the primary reason for their participation is the hope for cure or remission of their condition.[43]

We see once again, as with informed consent in the physician-patient relationship, that when society deems a major paradigm shift in the relationship between medical investigator and research subject to be essential, it is law, not medical ethics, to which it turns for realistic solutions. While ideally we might have hoped for the initiation of a national dialogue in which all of the stakeholders have participated in a good faith effort to arrive at a new social consensus, the effort by the Congress, spearheaded by Senator Walter Mondale, to bring that about was openly and vehemently opposed by the leading figures in medical research, as David Rothman has so carefully chronicled.[44]

When we move, as we shall in Chapter 5, from respect for patient autonomy to respect for patient prospective autonomy in the form of advance directives, both patients' rights advocates and the medical profession looked to state legislatures for certain reassurances with regard to rights and responsibilities respectively. Such reassurances, presumably, could not have been provided, at least with sufficient particularity and authority, through some type of ethical consensus on patient rights and physician responsibilities. The imprimatur of the law was deemed by most everyone to be essential.

ENDNOTES AND REFERENCES

1. Ludwig Edelstein, "The Hippocratic Oath: Text, Translation and Interpretation," *Bulletin of the History of Medicine* (Supp. 1, 1943).
2. Roe v. Wade, 410 U.S. 113, 130-132 (1973).
3. Examples include John Gregory, *Lectures on the Duties and Qualifications of a Physician* (London: Strahan & Cadell, 1772); Thomas Percival, *Medical Ethics* (Manchester: S. Russell, 1803); American Medical Association, "Code of Medical Ethics," in *Proceedings of the National Medical Conventions in New York, May 1846, and Philadelphia, May 1847.*
4. *Black's Law Dictionary* (St. Paul, MN: West Publishing Co., 1996).
5. Ruth R. Faden, Tom Beauchamp, Nancy M.P. King, *A History and Theory of Informed Consent* (New York: Oxford University Press, 1986), p.119.
6. 104 N.W. 12 (1905).
7. 104 N.W. 12, 14-15 (1905), emphasis added.
8. John Locke, *Second Treatise of Government* (N.Y.: Mentor Books, 1960), Ch. II, § 4.
9. Charles Fried, Correspondence, *Philosophy and Public Affairs* (1977), 6: 288-289.
10. 79 N.E. 562 (1906).
11. This expansive view of the prerogatives of the physician seems to have been both prevalent and persistent in the medical profession. In cases which were litigated long after informed consent became a fundamental principle of medical jurisprudence, similar assertions were made by physicians. *See, e.g.,* U.S. v. George, 239 F. Supp. 752 (1965) and Bouvia v. Superior Court, 179 Cal.App.3d 1127 (1986).
12. 105 N.E. 92 (1914).
13. 105 N.E. 92, 93 (1914).
14. Faden, et al., note 5, p. 123.
15. Salgo v. Leland Stanford Jr. University Board of Trustees, et al., 317 P.2d 170, 181 (1957).
16. Salgo, note 15.
17. 350 P.2d 1093 (1960).
18. 350 P.2d 1093, 1099 (1960).
19. 350 P.2d 1093, 1102 (1960).
20. 350 P.2d 1093, 1104 (1960).
21. 350 P.2d 1093, 1106-1107 (1960).
22. 350 P.2d 1093, 1106 (1960).
23. 464 F.2d 772, 776, 784 (D.C. Cir. 1972).
24. Jay Katz, *The Silent World of Doctor and Patient* (New York: Free Press, 1984), p. 2.
25. 464 F.2d 772, 787.
26. Carl F. Schneider, "Concluding Thoughts: Bioethics in the Language of the Law," in Carl F. Schneider, ed., *Law at the End of Life: The Supreme Court and Assisted Suicide* (Ann Arbor: University of Michigan Press, 2000), p. 298.

27. Nicholas Christakis, *Death Foretold: Prophecy and Prognosis in Medical Care* (Chicago: University of Chicago Press, 2000), Ch. 4.
28. American Medical Association, *Code of Medical Ethics*, (Chicago: American Medical Association, 1997), p. xxxix.
29. American Medical Association, *Code of Ethics*, (1947), Appendix A in Katz, note 24.
30. American Medical Association, *Code of Medical Ethics*, note 28, 8.08, p. 120.
31. For an historical analysis of this socio-political phenomenon *see generally* David J. Rothman, *Strangers at the Bedside – A History of How Law and Bioethics Transformed Medical Decision Making* (New York: Basic Books, 1991).
32. Faden, Beauchamp & King, note 5, pp. 141-142.
33. Faden, Beauchamp & King, note 5, p. 143.
34. California Medical Board, Action Report, Vol. 51, pp. 1, 8-9 (Oct. 1994); California Health and Safety Code, sect. 1, part 4.5 (1997).
35. Bergman v. Wing Chin, M.D., et al., Superior Court of California, Alameda County, No. H205732-1 (1999).
36. Evelyne Shuster, "Fifty Years Later: The Significance of the Nuremberg Code," *New Eng. J. Med.* (1997), 337: 1436-1440.
37. Henry K. Beecher, "Ethics and Clinical Research," *New Eng. J. Med.* (1966), 274: 1354-60.
38. David J. Rothman, note 31, pp. 73-75.
39. J. Jones, *Bad Blood* (N.Y.: Free Press, 1981).
40. 45 C.F.R. § 46.103 (1993).
41. 45 C.F.R. § 46.111 (1993).
42. 45 C.F.R. § 46.116 (a) & (b).
43. C. K. Daugherty, et al., "Perceptions of Cancer Patients and Their Physicians Involved in Phase I Trials," *Journal of Clinical Oncology* (1995), Vol. 13: 1062-1072..
44. David J. Rothman, note 31, pp. 168-189.

Chapter 5

FROM AUTONOMY TO PROSPECTIVE AUTONOMY: ADVANCE DIRECTIVES IN BIOETHICS, LAW AND PUBLIC POLICY

1. INTRODUCTION

Chapter 4 describes the pre-eminent role that the judiciary played in elevating respect for individual autonomy and informed decision making to a fundamental principle governing the physician-patient relationship. Nevertheless, many physicians, and especially those who began their clinical practice when the physician-patient paradigm was still the paternalistic one, challenged the doctrine of informed consent on both a theoretical and practical level. The essence of both levels of critique was that the typical patient did not have the capacity to evaluate data about diagnosis, prognosis, and alternative treatments so as to arrive at a reasonable conclusion about their medical care. One can well imagine, therefore, the heightened level of skepticism which they maintained when the focus shifted from the real-time decisions involved in informed consent and as an exercise of actual autonomy to the future, hypothetical decisions involved in advance directives as an exercise of prospective autonomy. The challenge to the moral and legal authority of advance directives as a legitimate exercise of a individual's right to prospective autonomy has come not only from the ranks of physicians, but from some philosophers and legal scholars as well. That critique is focused on a purported distinction between the identity of the person who executes the advance directive, let us call her the prospective patient, and the identity of the person whose treatment will be affected by the honoring of the directive, let us call her the actual patient. However, before exploring the philosophical issue in more detail, we need to be clear about what we mean when we use the term "advance directive."

2. THE NATURE OF ADVANCE DIRECTIVES

In the most elementary sense, an advance directive is a declaration (oral or written) by a person with decisional capacity of the types of medical treatment which they would or would not wish to receive during some future period of decisional incapacity. Originally, such directives were oral and amounted to nothing more than vague expressions of discontent over the prospect of "being kept alive only by machines" or having one's life sustained "as a vegetable." Based upon a demonstrated unwillingness of physicians and courts to honor such vague and informal expressions, a movement arose in the 1970's for the formal, legal recognition of such directives, initially in the form of a living will. With amazing rapidity, state legislatures jumped on the bandwagon and enacted living will statutes. The typical living will statute provides that a person with decisional capacity may execute a written document (living will) expressing the wish that life-sustaining interventions be withheld or withdrawn when they have lost decisional capacity and suffer from a terminal condition. Such statutes provide immunity from liability for physicians who respect the terms of the living will.

One of the more significant ironies about living will statutes is that although the generally strong public support for them arose out of a desire to avoid the fate of patients like Karen Ann Quinlan, whose life in a persistent vegetative state (PVS) was sustained for years through the provision of artificial nutrition and hydration, in fact most statutes were so narrowly drafted that they would not apply in Quinlan-type cases.[1] For example, the typical living will statute requires that the patient be certified by one or more (usually two) physicians to be in a terminal condition, and many statutes do not allow patients to prospectively refuse artificial nutrition and hydration as a form of life-sustaining treatment. Many physicians do not consider a PVS to be a terminal condition because with proper support such patients can live for many years. Moreover, most PVS patients require no invasive intervention other than artificial nutrition and hydration in order to be sustained. If that intervention cannot be prospectively declined, then there is no means available to avoid the fate of Karen Quinlan, who was kept alive for years in a permanently unconscious state.

In part because of the need for an advance directive that was free of the living will's most severe restrictions, another type has become quite common – the durable power of attorney for health care or health care proxy. The durable power of attorney allows an individual to designate in writing a proxy or surrogate decision maker (the attorney-in-fact) who has the same degree of authority to consent to or decline life-sustaining treatment as the patient would if he or she were competent. The expectation is that the proxy will have a sufficient understanding of what the wishes of the patient would be with regard to treatment during a period of decisional incapacity so that they can provide a substituted judgment for that of the decisionally-incapacitated patient. Regrettably, that expectation is not always reasonable because the typical durable power form requires neither a detailed specification of the patient's wishes or values nor confirmation that the patient has discussed those matters with their designated proxy.

With this cursory background on the nature of advance directives, let us return to the philosophical critique of them. Underlying the moral and legal authority of an advance directive is the basic presupposition that the person who executes the directive and the patient for whom the directive ultimately specifies the parameters of medical treatment are the *same* person. It is this presupposition which the philosophical critics of advance directives challenge. The critics, following the theory of personal identity most prominently advocated by the British philosopher Derek Parfit, assert that what makes a person at time T1 the same person as at time T2 is a minimally sufficient level of psychological continuity. In a person with normal mental faculties, there is a high level of continuity among psychological states even over many years. It is this continuity or consistency of memories, character traits, values, attitudes and the like, which causes us to think of and relate to someone over time as a particular person with a unique identity. We are speaking here, of course, about persons who are conscious, since the unconscious have lost (temporarily or permanently) the capacity to experience any degree of psychological continuity.

The quintessential cases in which issues of personal identity arise are those involving dementia. As the dementia progresses, the person is able to recall less and less about their life and their relationships with others, until finally they completely lose any sense of personal identity and any capacity to recognize even their closest relatives and friends. The severely demented person, according to Parfit, cannot possibly be the same person as their former competent self because a minimally sufficient level of psychological continuity no longer exists. Those who espouse the Parfitian view of personal identity maintain that an advance directive necessarily lacks any moral authority to dictate treatment or nontreatment of such a demented patient because the person who executed it no longer exists.

While there are many problems with the use of and adherence to advance directives in the health care setting, they are not a result of a broad acceptance of the Parfitian view of personal identity. Indeed, many people would maintain, without any deep philosophical conviction, that physical rather than psychological criteria should be dispositive in questions of personal identity. In other words, if we are talking about the same physical body, then we are talking about the same person. Even if one were to concede that psychological factors are important to considerations of personal identity, that does not logically entail the conclusion that advance directives lack moral authority in the case of the permanently demented. The primary concern about treatment decisions for patients who lack the capacity to decide for themselves is who should have the authority to make them. Those who contest the moral authority of advance directives generally argue that some third party should make such decisions based upon an objective assessment of the patient's best interests. Those who defend the moral authority of advance directives often take the position that the person executing the directive is in a superior position to make such determinations because it is the care of their own later demented selves that is at issue.

3. THE COURTS AND ORAL DIRECTIVES

Most state advance directive statutes declare that they are not intended to revoke or otherwise limit the common law rights of persons to direct their medical care.[2] In recognition of such consistently expressed legislative intent, the majority view among state courts has been that the existence of statutorily recognized forms for written medical directives should not undermine the authority of unambiguous and uncontested oral directives about medical care expressed by individuals at a time when they had decisional capacity. Indeed, even when oral directives are lacking, most courts do not presume that the default position should be unlimited treatment. Rather, in the absence of oral or written directives by the patients, the courts look to the objective "best interests" of the patient. While medical testimony will certainly be considered in making such determinations, it will not be conclusive, nor necessarily any more influential than lay testimony by those who knew the patient best. A few courts, however, in what often become high profile cases, have demonstrated a disturbing tendency to discount or even completely disregard unambiguous oral directives. Let us consider one of the most egregious examples, *In re Martin*,[3] a case in which a narrow and rigid view of the law was impervious to more enlightened bioethical analysis.

Michael Martin was a middle-aged husband and father of three children when he sustained severely debilitating injuries in an auto accident. As a result of those injuries, he permanently lost decisional capacity and the ability to do anything for himself. In addition to being unable to walk or talk, he required a colostomy for elimination of solid waste and a gastrostomy tube for artificial nutrition and hydration. As the courts subsequently took great pains to emphasize, he neither suffered from a terminal condition nor was he in a persistent vegetative state. Rather than being unconscious (comatose) or in a

PVS, such patients are described as in a "minimally conscious state." Whereas there is a strong consensus in the medical community that neither unconscious nor PVS patients have the capacity to experience pain and suffering, there is no such consensus with regard to minimally conscious patients. After spending time at several different nursing homes, Michael finally came to reside at a special facility for patients with severe neurological deficits. Five years after the accident, Mary Martin, Michael's wife of twenty years and his legal guardian and conservator, asked the bioethics committee of a hospital to which he had been admitted for treatment of a bowel obstruction whether life-sustaining medical treatment should be withdrawn. The committee issued a report concluding that "from an ethical standpoint, it was a reasonable request to discontinue nutritive support" and that such a request "was in accord with Mr. Martin's wishes as expressed to his wife and one of his co-workers." Nevertheless, the hospital declined to carry out the request to discontinue artificial nutrition and hydration without a court order.[4]

When Mary Martin filed the petition in probate court seeking that authorization, Michael's mother and sister filed a response in which both opposed the withdrawal of life support and requested that Mary be removed as Michael's guardian and conservator. During an evidentiary hearing lasting over two weeks, the petitioner presented testimony in support of her contention that Michael had made it clear on many occasions that he would not wish to be kept alive in such a severely debilitated condition. Mary Martin testified that Michael had always been very disturbed by persons who were disabled or dependent on others for their most basic needs, frequently indicating that he would rather die than be dependent on people and/or machines. In testimony and a supporting affidavit she noted that over a period of eight years before the accident the two of them had numerous conversations about their wishes regarding medical treatment if accident, disease, or old age should severely disable them. Michael repeatedly insisted that he would never wish to live in a state of total dependency, and actually told Mary that if she were to put him on a machine to keep him alive when he was irreversibly disabled and dependent on others, "I'll always haunt you Mary. Do you understand?"[5] Two co-workers of Michael's testified that he had told them shortly before the accident that he would not wish to be kept alive in a vegetative state. Michael's sister, one of the respondents in the case, acknowledged that he once told her he would not wish to be kept alive on a respirator if he were in a coma.

The respondents neither directly contested the truthfulness of the petitioner or Michael's co-workers, nor did they offer evidence of statements by Michael suggesting that he would find his current situation acceptable. They also did not contend that sustaining Michael's life in his current condition was in his best interests, perhaps because even the court-appointed guardian *ad litem* for Michael supported the wife's request that nutrition and hydration be discontinued. Their opposition seemed to be based upon their own religious or moral views about withdrawing treatment from someone whose life could be sustained.

The medical evidence concerning the level of Michael's cognitive functioning, the extent to which he could understand what was going on around him, and his capacity to experience pleasure or pain, was equivocal. All that the expert witnesses could agree upon was that he was conscious (as opposed to comatose or vegetative) and that his impairments were irreversible.

The probate court ruled that there was clear and convincing evidence that Michael would not wish to have his life sustained by medical means in his present condition, but that his wishes could not be respected because they had not been expressed in writing. The court denied the respondents' request to remove Mary Martin as guardian and conservator. The Court of Appeals reversed the probate court on the issue of the necessity of written proof of a patient's wishes, and remanded the case for further and more specific findings on certain evidentiary matters. The probate court arrived at the same conclusion as before that there was clear and convincing evidence of Michael's desire not to be sustained in his present condition.[6] Applying the usual standard of appellate review of civil trial proceedings, the Court of Appeals held that the trial court's determination – as to the weight of the evidence offered by the petitioner and the reasonable conclusions to be drawn from it – should be sustained because not clearly erroneous. Clearly, in the process of remand the Court of Appeals had disabused the trial court of the idea that Michigan law precluded the following of oral directives. The Court of Appeals also carefully noted that competent and credible evidence in the record supported the conclusion that Michael would not wish to have his life sustained if he were incapable of performing basic activities of daily living such as walking, talking, dressing, bathing, or eating. This, the court noted, is exactly the condition in which he now finds himself. The respondents appealed to the Michigan Supreme Court.

In its reversal of the Probate Court and the Court of Appeals, the Michigan Supreme Court did two things. First, as the dissenting opinion in the case points out, the majority opinion engaged in a *de novo* review of the evidence at trial, and in doing so arrived at the conclusion that it was not clear and convincing. Nowhere in that opinion is there a ruling, as one would expect, that the trial court's findings of fact were "clearly erroneous." Rather, the court de-emphasized the testimony of the petitioner about numerous and extensive conversations with her husband on the subject of living with disability, and focused almost exclusively on the fact that his statements to co-workers repeatedly referred to his unwillingness to live as a vegetable. By doing so, the court sought to justify its ultimate conclusion that the totality of the evidence failed to clearly and convincingly indicate Michael's desire to die rather than to continue living as one who is profoundly disabled and dependent.[7]

The irony of the insistence by the Michigan Supreme Court on absolute precision in advance directives declining treatment is that at the time of its *Martin* decision the Michigan legislature had failed to enact any form of advance directive statute.

The second thing the court did was to stake out its minority position on an objective "best interests" standard. Thus, Michigan is one of only a few states that insist that life-sustaining interventions be continued in the absence of clear and convincing evidence that the patient would not wish them, at least when the patient is not permanently unconscious.[8] By precluding any objective consideration of the best interests of such patients, these minority courts implicitly deny the possibility of what I refer to as "a medical fate worse than death." Permanently unconscious patients lack the capacity to experience such a fate, but those like Michael Martin may not. As a dissenting judge on the Missouri Supreme Court stated with regard to Nancy Cruzan: "If she has any awareness of her surroundings, her life must be a living hell."[9] The Michigan Supreme Court dismisses the possibility that Michael might have just enough awareness to realize his profound disability – that he is living just the fate he begged his wife to spare him – but not enough cognitive capacity to communicate his distress to those around him.[10] As Justice Brennan observed in his dissenting opinion in *Cruzan*, erring on the side of life, as some courts are wont to do, is by no means a risk-free proposition, for "an erroneous decision not to terminate life support . . . robs a patient of the very qualities protected by the right to avoid medical treatment. His own degraded existence is perpetuated; his family's suffering is protracted; the memory he leaves behind more and more distorted."[11]

The Martin decision appeared to present an ideal opportunity for the U.S. Supreme Court to further elaborate upon and explain a numberof issues which it only alluded to in passing in *Cruzan*, such as whether a state has a constitutional obligation to "defer to the decision of a surrogate if competent and probative evidence established that the patient herself had expressed a desire that the decision to terminate life-sustaining treatment be made for her by that individual."[12] Nevertheless, the petition for a writ of certiorari filed by Michael's guardian was denied.

The majority of courts recognize the many reasons why people do not execute advance directives, even those like Michael Martin who have strong feelings about continued existence in a severely disabled condition. Consequently, they do not insist upon written documentation of those wishes or extensive clinical detail about precisely which conditions or interventions the individual finds unacceptable.[13] More importantly, that majority eschews assuming a vitalistic default position, one which denies that there can be a medical fate worse than death. Instead, most courts recognize that when individuals like Michael Martin beseech their spouses and close relatives to protect them against the prolongation of a life which they (if they could still speak for themselves) would consider to be of unacceptably low quality, they are exercising a right that goes to the very "heart of liberty . . . the right to define one's own concept of existence."[14] Concerns among patients and their advocates that oral directives were not being appropriately considered by courts helped to generate the political momentum for the statutory recognition of advance directives.

4. THE LIVING WILL

The first generation of advance directive legislation was that of the living will. Such a document was intended to provide individuals with a statutorily recognized means of expressing in writing their desire that life-sustaining measures not be initiated or continued if they had lost decisional capacity and suffered from a medical condition which was terminal and irreversible. The political pressure exerted upon state legislatures to adopt such statutes came primarily from patient advocacy groups. They sought a document that would insure – presumably because its validity was recognized as a matter of law – that a patient's wishes would be followed so long as the prerequisite conditions had been met. However, a salient feature of these statutes was that they insulated physicians from liability for the fatal consequences of complying with a patient's expressed wishes to have life-sustaining measures withheld or withdrawn. As we shall see, immunizing physicians from liability was not a sufficient incentive for them to embrace living wills as a reasonable means of preventing the provision of unwanted medical measures at the end of life.

All living will statutes require that at least two conditions be confirmed before a valid document will become operative. The first is that the patient lacks decisional capacity. This determination can, in most states, be made by the attending physician and need not be confirmed by a consulting specialist, e.g., psychiatrist. Implicit in such a determination is that the patient is not likely to regain decisional capacity in the near future, i.e., the loss of decisional capacity is associated with the patient's underlying medical condition and will persist along with it. Patients whose inability to make decisions is episodic, and who thus have lucid periods in which they can communicate their wishes regarding treatment, do not have operative living wills.

The second and more troublesome requirement is that the attending physician, and usually a second consulting physician, must certify that the patient is in a terminal condition. Generally, a terminal condition is considered to be one that will result in the patient's death within six months. Many statutes, however, provide a more complex definition of "terminal." The Uniform Rights of the Terminally Ill Act, for example, defines a terminal condition as "an incurable and irreversible condition that, without the administration of life-sustaining treatment will . . . result in death within a relatively short time."[15] Life-sustaining treatment is defined by the Uniform Act as "any medical procedure or intervention that, when administered to a qualified patient, will serve only to prolong the process of dying."[16] In order to appreciate why living wills have not provided many patients with the deliverance from the onslaught of medical technology which they sought we need to consider these definitions in the context of typical cases.

The case of Karen Ann Quinlan arose in the early stages of the development of living wills. Quinlan was a 22-year-old woman who had been brought to a New Jersey hospital in April of 1975, in a state of unconsciousness apparently induced by ingestion of a combination of drugs and alcohol. Over the next several months, Karen showed no improvement, remaining in a PVS. She was maintained on a ventilator and received artificial nutrition and hydration. When her parents reluctantly arrived at the conclusion that her condition was irreversible, they requested that mechanical ventilation be withdrawn. The attending physician and the hospital declined to do so on moral grounds, forcing the Quinlans to retain an attorney and pursue the case all the way to the New Jersey Supreme Court. That court ultimately upheld the right of the Quinlans to direct the discontinuation of life-sustaining treatment for their daughter.[17] The attending physician did not, however, simply disconnect the ventilator, but very gradually weaned Quinlan from it. As a result, she did not die but continued to breathe on her own. Quinlan was then transferred to a long-term care facility where she survived for another ten years because her parents were unwilling to direct the cessation of artificial nutrition and hydration. Ultimately she died from pneumonia when her parents refused to consent to the administration of the standard antibiotic treatment.[18]

Although the Quinlan case did not involve a living will, it points out the problems that they pose. While she would never regain consciousness, she would not be deemed by some physicians to suffer from a terminal condition. Life-support – at least mechanical ventilation – was not merely prolonging the process of her dying, since she could be successfully weaned from the ventilator. While the discontinuation of artificial nutrition and hydration would have resulted in her death, her parents did not consider that to be life-sustaining treatment but rather food and water. For similar reasons, in fact, a number of the early living will statutes specifically excluded artificial nutrition and hydration as a life-sustaining medical treatment that could be declined through such an instrument.[19] It is not clear to what extent these exceptions were based upon a misinformed belief that unconscious patients can experience hunger and thirst, and to what extent they simply reflected the idea that providing food and water by any means is considered to be something other than medical treatment.[20] More recently, quite a few states have amended their statutes so as to allow artificial nutrition and hydration to be specifically declined by patients. This is consistent with the view of the U.S. Supreme Court expressed in *Cruzan v. Director, Missouri Department of Health*[21] that artificial nutrition and hydration, particularly when it is provided through a gastrostomy tube surgically implanted in the patient's stomach, constitutes a medical intervention and not merely the provision of food and water.

The bitter irony is that the very clinical cases which, by virtue of the significant media attention they received, galvanized public support for living wills, would not have been governed by such documents even if Karen Quinlan and Nancy Cruzan had executed them because of the terminal condition requirement and the artificial nutrition and hydration exception. If public concerns about being kept alive as a "vegetable" were going to be adequately addressed, public policy would have to move beyond the severely constrained device of the living will.

5. THE DURABLE POWER OF ATTORNEY FOR HEALTH CARE

For reasons that remain rather elusive, the only other form of advance directive to be recognized by statute, the durable power of attorney for health care (DPAHC), does not typically contain the restrictions found in living will statutes. Both forms of advance directive flow from the premise that respect for a person's right to exercise prospective autonomy should be commensurate with that same person's right to exercise actual autonomy. Nevertheless, as we have seen, the exercise of prospective autonomy through the living will is severely constrained in ways that would be unacceptable in the case of a competent patient refusing treatment at that moment. One possible explanation for the asymmetry is the view that persons lack the ability to adequately anticipate how they will feel about medical interventions for some future condition, particularly if they are neither disabled nor seriously ill at the time they execute the directive. Consequently, they must somehow be protected from their own potentially rash judgements about quality of life and ability to bear life-limiting conditions. There is, then, a strange sort of Catch 22, oxymoronish character to living wills. What appears to have been given with the right hand of the legislature has been subtly taken away with the left.

The same critique cannot be leveled at most DPAHC statutes. The health care proxy is empowered in most states to exercise the same degree of authority to give or withhold consent to any and all medical interventions as the patient would have if he or she had decisional capacity. The most common exception is a requirement of a terminal condition or permanent coma (or persistent vegetative state) in order to withhold or withdraw certain life-sustaining interventions.[22] One possible reason why the majority of DPAHC statutes allow the designated proxy to exercise the full range of decisionmaking authority that the patient would otherwise have is that more traditional forms of the power of attorney allowed the designated attorney-in-fact to exercise all of the powers necessary to manage the affairs of the person executing the document unless otherwise restricted. Another reason, although not presently supported by the data, is a presupposition that in the process of deciding about, crafting, and executing a DPAHC, an individual will select a proxy who is fully informed about the values, goals, and priorities of the patient, and who therefore can reasonably be expected to make healthcare decisions that are consistent with those the patient would have made.[23]

The fact remains, however, that execution of a valid DPAHC requires neither a written articulation of the person's values, goals or priorities for medical care, nor certification that the same has been communicated to the health care proxy pursuant to the execution of the directive. Indeed, some commentators have advised against advance directives that specify particular conditions in which one would not wish to be maintained or particular interventions that one does not wish to receive.[24] Two common rationales offered for this advice are: 1) there may be an implication that one wishes everything not specifically mentioned, and 2) such specifications may unduly constrain the health care proxy when other unanticipated eventualities arise. While the typical DPAHC form provides no guidance on how to judiciously provide particulars or special considerations, there are precautions that can be taken so as to maximize the benefits and minimize the risks of adding substance to the skeletal DPAHC form. One approach would be to begin and conclude any discussion of specific conditions (e.g., PVS) or medical procedures (e.g., intubation and mechanical ventilation) of particular concern to the person executing the directive with the well-known legal phraseology that such details are provided "by way of example and not limitation." To make the point so that no one can misunderstand it, the DPAHC might then further provide that "my health care proxy is authorized to use his/her best judgment in any and all circumstances not obviously covered by the examples offered in this document." This point will be considered further in Section 9.

6. THE PATIENT SELF-DETERMINATION ACT

Despite the much greater ability of persons (at least in theory) to exercise their prospective autonomy through the execution of a DPAHC, the fact remains that significantly less than 20% of the general population of adults in the United States have executed a written advance directive.[25] One significant contributing factor is that the attitude of much of the medical profession to advance care planning generally and to advance directives in particular ranges from indifference to open hostility.[26] In political terms, this overwhelmingly negative attitude was translated into a well-organized and vociferous opposition to legislation sponsored by Senator John Danforth of Missouri entitled "The Patient Self-Determination Act" (PSDA). Indeed, both the American Medical Association and the American Hospital Association lobbied heavily, but ultimately unsuccessfully, against the bill. In principle, the PSDA constitutes a strong affirmation, as a matter of public policy, of the exercise of the individual right of prospective autonomy. Despite the implications of its title, the PSDA does not create a federal right to engage in advance care planning through the execution of advance directives. Operating on the presupposition that the small number of persons with a written advance directive is a consequence of a general lack of information about them, the PSDA creates an obligation on health care institutions, as a condition of their participation in the Medicare or Medicaid programs, to provide each patient with written information about his or her rights under state law (statutory or decisional) to make health care decisions (including refusal of treatment) and to formulate advance directives. The patient must also be provided with information about any written policies of the institution regarding the implementation of those rights.[27]

In theory, the penalty imposed upon an institution that fails or refuses to meet its responsibilities under the PSDA is draconian – mandatory loss of all Medicare and Medicaid funding with no exceptions and no extensions in order to come into compliance with the law. In practice, however, in the ten years since the PSDA became law, no institution has ever lost, or even been at serious risk of losing, its funding because of noncompliance with the PSDA. As we shall see in the next section, the absence of any PSDA enforcement action by the Department of Health and Human Services (HHS) in no way indicates universal compliance with its requirements. Rather, it suggests that senior administrators in HHS consider the punishment under the PSDA to be so disproportionate to the "crime" that justice and prudence warrant overlooking infractions and deficiencies.

In 1995, the General Accounting Office (GAO) produced a "progress" report on the implementation of the PSDA, in which a number of barriers to the realization of the legislative intent of the Act are identified. Remarkably, despite the previously noted fact that the PSDA places the entire responsibility for information disclosure and patient education on health care institutions, all of the barriers identified in the GAO report relate to physicians.[28] The barriers are reported to be: 1) lack of physician knowledge about advance directives; 2) physician belief that advance directives are not necessary for most patients; 3) the lack of reimbursement for physician time and effort in discussing advance directives with patients; and 4) widespread physician belief that because death is not an appropriate outcome of medical care, discussion of a patient's right to refuse life-sustaining treatment is not their responsibility.

This disconnect between what the PSDA provides and what the GAO considers to be the reasons why the objectives of the PSDA have not been realized is truly remarkable. Moreover, it brings into sharp focus what in fact is a fundamental, and perhaps a fatal flaw of the PSDA, which is that as a matter of legal obligation it removes advance directives and advance care planning from the physician-patient relationship. Under the PSDA and most, if not all, state advance directive legislation, physicians can cultivate their ignorance of a patient's advance directives and attitudes toward life-sustaining treatment with impunity and at no legal risk. For example, although all states now have some statutorily-recognized process for creating advance directives, less than one-third have any sanctions against physicians for failure to follow a valid directive or transfer the care of the patient. Of those with a provision for sanctions, only six impose some form of civil or criminal liability. The remaining statutes indicate that such failure may be deemed to constitute unprofessional conduct. However, disciplinary actions for this form of unprofessional conduct, like the imposition of sanctions by HHS for failure to comply with the PSDA, are nonexistent.

7. SUPPORT AND ITS ETHICAL IMPLICATIONS

The stated purpose of the Study to Understand Prognosis and Risks of Treatment (SUPPORT) was to "improve end-of-life decision making and reduce the frequency of a mechanically supported, painful, and prolonged dying process."[29] The study, conducted in five elite American teaching hospitals, consisted of two phases: a two-year prospective observational study and a two-year controlled clinical trial. The results of the first phase revealed a significant disparity between the expressed wishes of patients or proxy decision makers with regard to cardio-pulmonary resuscitation (CPR) and the entry of do-not-resuscitate (DNR) orders, life-sustaining interventions such as mechanical ventilation, artificial nutrition and hydration, and pain and symptom management. Although all of the patients in the study were in an intensive care unit (ICU) and at significant risk of death in the hospital, the investigators found that discussions about end-of-life care and decision making between physicians, patients, and family were "uncommon."

The second phase of the study was designed to improve physician-patient-family communication and decision making regarding the care of gravely ill patients by "providing timely and reliable prognostic information, eliciting and documenting patient and family preferences and understanding of disease prognosis and treatment, and by providing a skilled nurse to help carry out the needed discussions, convene the meetings, and bring to bear the relevant information."[30]

When the data from the second phase had been fully analyzed, the intervention was found to have had no measurable impact on the quality of communication or respect for patient wishes. More particularly, the investigators found that:

> . . . even though the targeted outcomes are objectives of much ethical and legal writing and of some explicit social policy (such as informed consent statutes, the Patient Self-Determination Act, and guidelines on pain), there were no secular trends toward improvement for intervention or control of patients during the 5 years of SUPPORT data collection.[31]

SUPPORT graphically illustrated and copiously documented the grim truth of the assertion by Yale surgeon Sherwin Nuland that despite all of the admonitions by ethicists and threats by patients' rights advocates, "One way or another, the rescue credo of high-tech medicine wins out, as it almost always does."[32]

What are the lessons of SUPPORT, and what are their implications for law and biomedical ethics? The SUPPORT Principle Investigators have openly and notoriously cringed at the "media hype" that followed in the wake of their report, especially the harsh criticism of both the competence and the motives of the physicians who care for gravely ill hospitalized patients.[33] To be sure, a close and objective reading of SUPPORT reveals, as one might reasonably expect, that there are many factors contributing to the perceived deficiencies in patient care. For example, while physicians were not, generally speaking, initiating discussions with ICU patients or their families about the goals of treatment, the likelihood they might be realized, and the role of invasive measures in achieving those goals, neither were patients anxious to have these discussions.[34] SUPPORT can be interpreted so as to provide support for the proposition that imminently dying ICU patients are being flogged with invasive procedures that, objectively assessed, have only a minimal likelihood of success. The data actually indicates, however, that according to the best medical prognostic instruments, on the day of death the median patient had one-to-four odds of living two months, and five days prior to death the median patient had a 40% likelihood of living two months.[35] With the gift of hindsight, one can marvel at the naiveté of the investigators in harboring the belief that a relatively simple and straightforward intervention – the introduction of a nurse-facilitator – would demonstrably improve the designated outcomes.

SUPPORT suggests, and its Principle Investigators confirm, that patients and families are complicit, passively if not actively, in the communication problems that beset the care of gravely ill and dying patients in the hospital. Consequently, admonitions such as that by health law expert George Annas that the best way to improve the care of dying patients in the short run is to get them out of the hospital, or to keep them from going to the hospital at the end of life, are not likely to be followed. Similarly implausible, at least in the current political environment, is his suggestion that we establish a system of not-for-profit public interest health care law firms with the role and mission of vindicating the legal rights of dying patients through a combination of public education and litigation.[36] If, as Annas concedes, "[p]hysicans simply have never taken the rights of hospitalized patients seriously," despite the threat of malpractice liability for treatment that lacks informed consent, yet another threat of litigation is unlikely to succeed where Phase II of SUPPORT failed, i.e., to induce physicians to engage in meaningful dialogue with patients about the curative and palliative goals of treatment.[37]

8. LEARNING THE LESSONS OF THE PSDA AND SUPPORT: A REFORM AGENDA

The primary objective of the PSDA was to encourage patients to execute advance directives to ensure that their end of life care would be consistent with their wishes and values. The primary objective of Phase II of SUPPORT was to maximize the extent to which the care of ICU patients was consistent with their own wishes and goals. While the PSDA placed the responsibility for discussing advance directives with patients on institutions, SUPPORT sought to change physician behavior through an intervention orchestrated by a nurse-facilitator. Thus, the PSDA implicitly suggested that physicians have no direct responsibility to engage in advance care planning with patients, while SUPPORT explicitly targeted the physician. Both efforts failed miserably, in part because of a significant discontinuity between the ethical ideals for patient autonomy our society espouses in theory, and a prevailing culture of medicine that is at best indifferent and at worst hostile to the realization of those ideals in most patient care settings.

Attempts at "moral suasion" directed toward physicians have failed as profoundly as the PSDA and SUPPORT. Prior to both, a distinguished group of physicians wrote in the New England Journal of Medicine: "Physicians have a responsibility to consider timely discussions with patients about life-sustaining treatment and terminal care . . . Such discussions and the resultant documentation should be considered part of the minimal standard of acceptable care." They went on to say: "To allow patients to experience unbearable pain or suffering is unethical medical practice."[38] We will consider aspects of this seminal article again in Section 10. If such admonitions had truly found their way into the culture of clinical practice, the physicians responsible for the patients whose care was the subject of SUPPORT might consider themselves to be at risk of disciplinary action and/or malpractice liability. Clearly, they did not. Moreover, with the exception of the previously noted remarks by George Annas, nowhere in the voluminous literature spawned by SUPPORT is there any suggestion of unprofessional conduct on the part of physicians.

It might be tempting to some to seize upon the SUPPORT scenario as a singular opportunity to engage in comparative bioethical analysis. Advocates of feminist, care, and narrative ethics can easily construct a plausible critique of the conduct of the attending physicians (regardless of their actual gender) for these ICU patients. It would not be pure caricature to characterize their behavior as a throwback to a previous era of hidebound physician paternalism, in which brazen indifference was shown to the patient's story, to the possibility that care might be more important than cure to such patients, or that medical chauvinism motivated the deaf ear that physicians turned to the nurse facilitators in the second phase of SUPPORT. One could also portray SUPPORT as a pitched battle between deontological and utilitarian factions. The deontologists would insist that ICU patients be treated as unique individuals who are ends in themselves, not merely the means by which elite academic physicians push the envelope in the high tech management of critical care. A utilitarian might respond that the life-extending interventions provided to the patients whose care was under scrutiny in SUPPORT furthered both professional and patient interests.

But nothing in the literature supports the proposition that such theoretical disputes credibly explain the findings of SUPPORT. Indeed, the most plausible explanation that has emerged has nothing whatsoever to do with doctrinal bioethical discourse. Rather, at the root of the problem are systems that support the development and implementation of medical interventions with little or no consideration of what the patient would wish to do if fully advised of their prognosis and the benefits and burdens of those interventions.[39] Systemic problems are likely to require systemic solutions. While bioethical theory may bolster the need for developing systemic changes, it is less likely to influence, at least in any significant way, the precise nature of those changes. An example may be found in the recently implemented accreditation standards of the Joint Commission for the Accreditation of Health Care Organizations

(JCAHO) regarding the management of pain. Among the data to which JCAHO was responding when it decided to develop these standards was the indication in SUPPORT that approximately 50% of all patients who died in the ICU suffered from moderate to severe pain in the last few days of their lives. The new JCAHO standards are intended to inform patients and their families of the right to the prompt and effective assessment and management of their pain, and to require accredited institutions to put in place systems that will insure that pain is regularly monitored (as are the patient's other vital signs) and effectively managed.

A new generation of advance directives, if they can be made part of the minimal standard of acceptable care, offers the promise of correcting some of the more egregious shortcomings of statutorily recognized advance directives, the PSDA, and the care of gravely ill patients as identified by SUPPORT. Use of these directives would be entirely consistent with the norms for physician conduct elucidated in the New England Journal article referred to at the beginning of this section. While it is true that advance directives do not, strictly speaking, direct the care of patients who have not lost decisional capacity, they can and should inform discussions about advance care planning, as well as the treatment of competent patients who are gravely ill.

9. THE MEDICAL DIRECTIVE

Characterized by its creators as a "comprehensive advance care planning document," the Medical Directive makes two important modifications of the earlier forms of advance directive.[40] First, it is designed to be completed through a collaborative effort between the patient and his or her personal physician, thereby placing advance care planning where it belongs, at the heart of the physician-patient relationship.[41] Second, it expresses the patient's wishes and treatment goals in terms of particular types of medical conditions and with regard to specific medical interventions.

The original version of the Medical Directive contained four hypothetical clinical scenarios or "situations" of varying severity. These range from brain damage that renders meaningful interaction with others impossible, but with no other co-morbidities, to permanent coma with another terminal condition such as an advanced stage of metastatic cancer. A subsequent version added a scenario to be specified by the patient in consultation with his or her physician, and a scenario based upon the individual's current health status but with a temporary loss of decisional capacity. For each of these scenarios, the individual is asked to indicate whether he or she would (if medically indicated) wish to receive any of the following interventions: cardiopulmonary resuscitation, mechanical ventilation, artificial nutrition and hydration, major surgery, kidney dialysis, chemotherapy, minor surgery, invasive diagnostic tests, blood transfusions, antibiotics, simple diagnostic tests, pain medications. The options with regard to each of the interventions include (if applicable): unqualified utilization, a trial with discontinuation if no clear improvement is forthcoming, nonutilization, or undecided.

The later version of the Medical Directive also asks the patient to indicate in each clinical scenario what the overall goals of the treatment or care plan should be: prolonging life by treating everything, attempting to cure but reevaluating often, limiting treatment to less invasive and burdensome interventions, comfort care only, or some other goal specified by the patient. This is a critically important addition because too often in the care of gravely ill hospitalized patients the decision making process is focused exclusively on procedures rather than on the ultimate goals of treatment and the likelihood that they can be realized. Among the insights gained from SUPPORT is the importance of physician-patient communication on the goals of care so that the treatment patients actually receive is "tailored to meet their goals and preferences."[42]

The Medical Directive is a "comprehensive advance care planning document" because, in addition to the matrix of clinical scenarios and detailed list of potential interventions, it also provides for the designation of a Durable Power of Attorney for Heath Care (DPAHC), organ donation, and a personal statement in which the individual can indicate which medical conditions would so significantly compromise quality of life that further life-sustaining measures should be discontinued. The inclusion of the DPAHC within the Medical Directive sends an important message that the mere designation of a health care proxy does not complete the advance care planning process. The DPAHC portion of the Medical Directive has been significantly expanded in the later version. Among the changes that have been made are the inclusion of a provision by which the individual indicates how conflicts between previously stated preferences and the views of the proxy are to be resolved. While we might expect most patients to indicate that their wishes should determine the treatment provided rather than those of the proxy, that may not be the case.[43] Another new provision is an optional acknowledgment by the personal physician that he or she has read the Medical Directive, discussed it with the patient, and acknowledges his or her duty to implement the stated preferences in order to fulfill the patient's wishes.

The organ donation provision in the Medical Directive is more important than many individuals may realize. It is not generally known by the lay public that despite the existence of an organ donor designation, such as on a driver's license, the custom and practice throughout the United States is to ask the deceased patient's family whether they will consent to organ donation. The family is under no legal (as opposed to ethical) obligation to do so, and if they decline, their wishes are uniformly honored, even though they are inconsistent with those of the would-be donor.[44] There is some chance, however slight, that by including the intent to donate organs in a Medical Directive, the family may be less inclined to override the patient's wishes. Furthermore, the designated surrogate should feel both a moral obligation and the legal empowerment to persuade the family of the importance of respecting the patient's wishes.

In designing the Medical Directive, with its range of clinical scenarios and detailed list of interventions, its creators were seeking to deal with the problem of vagueness that afflicts so many oral and written directives, including those that are statutorily recognized. However, erring on the side of detail is not without potential risks, which is why, as we have previously noted, some advocates of advance care planning nevertheless advise against the preparation of detailed directives.[45] A reasonable compromise between directives that can be dismissed as "hopelessly vague" and directives which are interpreted much too narrowly because they seem to cover only those situations or interventions that are specifically mentioned can be found with the Medical Directive as the framework. The introductory language in the Medical Directive provides that it shall "stand as a guide to my wishes regarding medical treatments in the event that illness should make me unable to communicate them directly." To this should be added language such as the following:

> The hypothetical situations and list of clinical interventions set forth in this document are provided **by way of example and not limitation**. My health care proxy, attending physician, as well as any court, shall make a good faith effort to apply the sense of this Medical Directive as a whole to my actual clinical situation. **Under no circumstances** should the listing of particular hypothetical scenarios or clinical interventions be interpreted so as to suggest that under all other circumstances I wish aggressive life-sustaining interventions to be pursued.

With this additional language, the Medical Directive is more likely to insure that only treatment consistent with the patient's values is provided than would a directive which simply designates a health care proxy, but gives no written guidance with regard to the values that are to guide the proxy in making decisions.

10. THE VALUES HISTORY

The Medical Directive, as previously noted, provides for a personal statement. That statement is intended to be an expression of the individual's values and priorities in the face of decisional incapacity and serious illness. During the last fifteen years such declarations have come to be referred to as a "values history."[46]

There are two ways in which to conceptualize the values history. The first is in the broader context of patient histories. An essential aspect of medical education and clinical practice is the taking of histories – the general medical history, family history, and history of present illness. In order to work effectively with patients in advance care planning, physicians must learn how to take yet another kind of history – a values history. Such a history appears to be precisely what Wanzer, et al., have in mind when they write:

> . . . as a matter of routine, physicians should become acquainted with their patients' personal values and wishes and should document them just as they document information about medical history, family history, and sociocultural background. Such discussions and the resultant documentation should be considered a part of the minimal standard of acceptable care. The physician should take the initiative in obtaining the documentation and should enter it into the medical record.[47]

So conceived, the values history, unlike other forms of advance directive, is actually an essential aspect of the physician's responsibility to generate information about the patient that is integral to good care and to integrate that information into the patient's medical records. Just as the other forms of patient history do not depend upon discussions initiated by patients, the same should be the case for the values history.

Among the questions a physician should ask in the course of taking a values history is whether the patient's preferences for treatment in the event of serious illness and decisional incapacity have been discussed with a close relative or friend. If so, then the nature and extent of those discussions should be carefully noted in the history. If those preferences have not been incorporated into a written directive, the values history presents an excellent opportunity for the physician to encourage the patient to do so. This is also an opportune time for the physician to suggest a follow-up meeting that would, if practicable, include the relative or friend with whom these discussions have taken place and who is the patient's designated proxy. The primary purpose of such a discussion would be to ascertain the level of congruence between the physician's understanding of the patient's wishes and that of the proxy.[48]

If the patient has not considered advance care planning, or discussed such preferences for treatment with another person, the physician can take this opportunity to explain the importance of doing so. As a means of facilitating the patient's reflection upon the role of personal values in advance care planning, the other conceptual model for the values history can be utilized. This model was developed at the Institute of Public Law at the University of New Mexico School of Law.[49] It contains an extensive list of thought-provoking questions, such as:

- What do you fear most? What frightens or upsets you?
- What, for you, makes life worth living?
- How do you expect friends, family, and others to support your decisions regarding medical treatment that you may need now or in the future?
- If you were to experience decreased physical and mental abilities, then how would that affect your attitude toward independence and self-sufficiency?
- How do your (religious/spiritual) beliefs affect your feelings toward serious, chronic, or terminal illness?
- What will be most important to you when you are dying (e.g., physical comfort, no pain, family members present, etc.)?
- What general comments would you like to make about your finances and the cost of health care?

The University of New Mexico Values History Form is not one that can be completed without serious reflection on the values that inform a person's life. Such reflection is not a common characteristic of Western culture at the beginning of the Twenty-First Century.[50] An interesting approach to inducing individuals to engage in thoughtful and deliberate advance care planning, necessitating, as it does, recognition of our vulnerability to disease and serious disability, as well as our ultimate mortality, has been taken by a group called Aging with Dignity. This group has developed an advance care planning document called *"Five Wishes,"* which are: 1) the person I want to make health care decisions for me when I can't make them for myself; 2) the kind of treatment I want or don't want; 3) how comfortable I want to be; 4) how I want people to treat me; and 5) what I want loved ones to know.[51] Page one of the form states: "*Five Wishes* is a gift to your family members and friends so that they won't have to guess what you want." Such a statement is not mere hyperbole, since in the absence of clear guidelines, decisions by proxies whether or not to continue life-sustaining medical interventions when the prognosis is poor but ultimately uncertain are likely to be anguishing.[52] While such decision making will always be difficult, indeed painful, a realistic basis for believing that the patient's wishes and values are being respected should eliminate a great deal of the unnecessary suffering that arises from lost opportunities for communication.

In concluding this chapter, it is advisable to review some of what we have learned from over a quarter century of experience with advance directives and other efforts to enhance respect for patient autonomy and dignity at the end of life. Efforts to encourage advance care planning which isolate physicians from the process and insulate them from any professional responsibility to become involved are doomed to failure. Indeed, efforts to encourage physicians to communicate and consult with gravely ill patients and their families about treatment goals, which was the objective of SUPPORT, will also fail if they do not address systemic problems.[53] It is difficult to believe that if the shared decision making model of patient care presupposed by the informed consent principle were now well-accepted and consistently practiced in other health care and hospital venues, it would be so patently disregarded in the ICU. Nor should too much emphasis be placed upon the uncertainties of prognostication for gravely ill patients. Such uncertainties are simply a component of the disclosures that physicians are obliged to make to patients or surrogates, not excuses for incomplete or nondisclosure.[54] The implication, therefore, is that fifty years of informed consent legal doctrine to the contrary notwithstanding, as well as the clear articulation of bioethical principles that are consistent with it, health care organizations and the professionals who work within them feel little obligation to design and carry out a plan of care that is consistent with the informed decisions of patients or their designated surrogates.[55] Systemic change may well be what is required. However, the unknown factor is what will be necessary to motivate such change and insure compliance with the new systems that are put in place. Both incentives for compliance and sanctions for noncompliance are likely to be required. As soon as one posits sanctions for noncompliance, the law looms large. Once again, the lessons of advance directive legislation, the PSDA, and SUPPORT are that sanctions must be immediate, significant, credible, and consistently imposed, otherwise they are not only ineffective, but actually engender a wholesale disregard for that with which they were intended to insure compliance.

ENDNOTES AND REFERENCES

1. Karen Quinlan was 21 years old when she was admitted to the intensive care unit of a New Jersey hospital after having lost consciousness and going into respiratory arrest. While the cause remained subject to dispute, it was clear that she had been dieting, and that she had consumed alcohol and ingested valium earlier in the evening. When it became clear after several months that she would never regain consciousness, her parents requested that she be taken off the respirator and be allowed to die. Her physicians declined to do so and a long legal battle ensued. For a detailed account of this battle *see* David J. Rothman, *Strangers at the Bedside* (N.Y.: Basic Books, 1991) and Gregory E. Pence, *Classic Cases in Medical Ethics*, 2nd ed. (N.Y.: McGraw-Hill, Inc., 1995).
2. The Colorado statute, for example, provides in part that: (f) It is the legislative intent that nothing in this article shall have the effect of modifying or changing currently practiced medical ethics or protocol with respect to any patient in the absence of a declaration as provided for in section 15-18-104; (g) It is the legislative intent that nothing in this article shall require any person to execute a declaration. C.R.S. § 15-18-102.
3. 538 N.W.2d 399 (Mich. 1995).
4. Petition for Writ of Certiorari to the U.S. Supreme Court, Martin v. Martin, Nov. 20, 1995, p. 3.
5. Petition, note 4, p. 6.
6. 517 N.W.2d 749 (1994).
7. In re Martin, 538 N.W.2d 399 (Mich. 1995).
8. Other states holding this minority view include New York and Missouri. As in Martin, both decisions by the highest court of those states engaged in the highly disfavored tactic of reversing the trial court's finding of clear and convincing evidence to support the withdrawal of life support by engaging in a *de novo* review of the evidence. *See* Cruzan v. Harmon, 760 S.W.2d 408 (Mo. 1988); In re Westchester County Medical Center, 531 N.E.2d 607 (1988).
9. 760 S.W.2d 408, 429 (1988).
10. One of the points on which the expert testimony was conflicting was whether Michael had the ability to understand questions as to whether he wished to go on living or to communicate by nods a meaningful response to them.
11. Cruzan v. Director, Missouri Dept. of Health, 110 S.Ct. 2841, 2873 (1990), Brennan, J. dissenting.
12. 110 S.Ct. 2841, 2856 (1990), n. 12
13. Justice O'Connor expressed precisely this point when she stated: "Few individuals provide explicit oral or written instructions regarding their intent to refuse medical treatment should they become incompetent. States which decline to consider any evidence other than such instructions may frequently fail to honor a patient's intent." 110 S.Ct. 2841, 2857 (1990), O'Connor, J. concurring.
14. Planned Parenthood of Southeastern Pennsylvania v. Casey, 505 U.S. 833, 851 (1992).
15. Uniform Rights of the Terminally Ill Act of 1989 § 1(9) 9 B U.L.A. 138 (Supp. 1994).
16. Uniform Rights of the Terminally Ill Act, note 15, § 1(4).
17. *In re Quinlan*, 355 A.2d 647 (N.J. 1976), cert. denied sub nom.
18. For a detailed account of the Quinlan case from the perspective of her parents *see* Joseph Quinlan, Julie Quinlan, *Karen Ann: The Quinlans Tell Their Story* (Garden City: Doubleday, 1977).
19. In the Cruzan case, one of the dissenting judges on the Missouri Supreme Court characterized the Missouri living will statute as a "fraud" perpetrated on the citizens of the state because of the many procedures, such as artificial nutrition and hydration, which cannot be declined. Cruzan v. Harmon, 760 S.W.2d 408, 442 (Mo. Banc 1988).

20. For conflicting views on the propriety of the withholding and withdrawing of artificial nutrition and hydration *see* Bonnie L. Steinbock, "Artificial Feeding: Solid Ground, Not a Slippery Slope." *New England Journal of Medicine* (1988), 318: 286-290; Mark Siegler, Alan Weisbard, "Against the Emerging Stream: Should Fluids and Nutritional Support Be Discontinued?" *Archives of Internal Medicine* (1985), 145: 129-131.
21. 110 S. Ct. 2841 (1990).
22. Eleven states have such restrictions. For an example of how they are articulated *see* Ohio Rev. Code Ann. § 1337.13. (B) (1993).
23. *But see* Nancy R. Zweibel, Christine K. Cassel, "Treatment Choices at the End of Life: A Comparison of Decisions by Older Patients and Their Physician-Selected Proxies," *Gerontologist* (1989), 29: 615-621.
24. Allan S. Brett, "Limitations of Listing Specific Medical Interventions in Advance Directives," *Journal of the American Medical Association* (1991), 266: 825-828.
25. American Medical Association Surveys of Physicians and Public Opinion on Health Care Issues, (1988), pp. 29-30.
26. One empirical study of physicians revealed outright antagonism toward advance directives based primarily on one of two perceptions: that they interfered with the exercise of professional judgment and/or that they called into question the physician's ability to determine what is in the patient's best interests. Joel M. Zinberg, "Decisions For Dying: An Empirical Study of Physicians' Responses to Advance Directives," *Vermont Law Review* (1989), 13: 445-491.
27. 42 U.S.C.A. § 1395cc(a)(1)(f)(1)(A).
28. It is probably beyond serious dispute that if the PSDA had assigned physicians rather than institutions with the responsibility to inform patients about their rights to consent to or to refuse treatment and to execute advance directives, the collective political power of organized medicine would have insured the defeat of the PSDA. Placing these responsibilities on institutions in effect insulated physicians from any significant legal responsibilities (at least under federal law) in the advance care planning process.
29. The SUPPORT Principal Investigators, "A Controlled Trial to Improve Care for Seriously Ill Hospitalized Patients," *Journal of the American Medical Association* (1995), 274: 1591-1598, 1591.
30. SUPPORT, note 29, p. 1595.
31. SUPPORT, note 29, p. 1595.
32. Sherwin B. Nuland, *How We Die* (New York: Alfred A. Knopf, 1994), p. 253.
33. Joanne Lynn, co-director of SUPPORT, observes that even the AMA appeared to have been caught up in the media frenzy to "scapegoat" physicians when it issued a press release with the following lead: "Care for Dying Americans Needs Substantial Improvement, Say Researchers in Largest-Ever Study of People Near Death: Study Finds Too Many Die Alone, in Pain, and Attached to Machine,." S. L. Isaacs, J. R. Knickman, eds., *To Improve Health and Health Care 1997 – The Robert Wood Johnson Anthology* (San Francisco: Jossey-Bass Publishers, 1997), p. 175.
34. For example, of those patients in the study who had not talked with a physician about CPR, only 42% indicated that they wished to do so. The remainder, a majority, reported a preference that family members and/or physicians make such decisions for them. Joanne Lynn, et al., "Preferences For Instructional Directives or Family Decision Making Among Older Patients," *Journal of the American Geriatric Society (*1994), 42: 217.
35. Joanne Lynn, Frank E. Harrell, Jr., Felicia Cohn, "Defining the Terminally Ill: Insights From SUPPORT," *Duquesne Law Review* (1996), 35: 311-336.
36. George J. Annas, "How We Lie," *Special Supplement, Hastings Center Report* (1995), 25: S12-S-14.

37. Annas suggests that [m]edical licensing boards must make it clear to licensees that painful deaths are presumptively ones that are incompetently managed and should result in license suspension or revocation in the absence of a satisfactory justification. Interestingly, following the Supreme Court decisions in the physician-assisted suicide (PAS) cases, the organization Compassion In Dying, spearheaded by its legal counsel Kathryn Tucker, who argued the PAS cases before the U.S. Supreme Court, has devoted considerable energy toward legal efforts to improve the care of dying patients. One of their most significant problems has been to identify patients or families who will come forward and report instances of substandard care at the end of life which can then be submitted, in the form of a complaint, to the appropriate state medical board for review.
38. Sidney Wanzer, et al., "The Physician's Responsibility Towards Hopelessly Ill Patients – A Second Look," *New England Journal of Medicine* (1989), 320: 844-849.
39. For early indications that SUPPORT points toward systemic problems *see* Bernard Lo, "End-of-Life Care After Termination of SUPPORT," *Special Supplement, Hastings Center Report* (1995), 25: S6-S8; Mildred Z. Solomon, "The Enormity of the Task: SUPPORT and Changing Practice," *Special Supplement, Hastings Center Report* (1995), 25: S28-S32.
40. Ezekiel Emanuel, Linda Emanuel, "The Medical Directive – A New Comprehensive Advance Care Planning Document," *Journal of the American Medical Association* (1989), 261: 3288-3295.
41. The immediate and consistent response to this sort of suggestion is that in the new, cost-conscious managed care environment there is simply no time for physicians to indulge in such conversations with their patients. If, in fact, a cost-benefit analysis has been done (a dubious assumption), the results of which suggest that such discussions are not worth the physician time involved, then I suggest it is flawed. The fruits of such discussions are usually an expression by patients that they do not wish that "everything" be done when their condition is grave and their prognosis grim. Yet SUPPORT and a mountain of anecdotal information confirm that in the absence of clear directives, costly interventions are continued far beyond the point at which they might reasonably have been withdrawn.
42. Joan Teno, et al., "Decision-making and Outcomes of Prolonged ICU Stays in Seriously Ill Patients," *Journal of the American Geriatrics Society* (2000), 48: S70-S74.
43. At least with regard to cardiopulmonary resuscitation, however, recent data suggests that a majority of patients prefer that their DNR status be determined at the relevant time by their physician and family/health care proxy rather than by their own previously expressed preferences. Christina M. Puchalski, et al., "Patients Who Want Their Family and Physician to Make Resuscitation Decisions For Them: Observations From SUPPORT and HELP," *Journal of the American Geriatrics Society* (2000), 48: S84-S90.
44. There is clearly a risk management element in such a policy. If the patient's closest relatives seek to override his decision to donate organs, there is no one who is likely to launch a legal challenge. On the other hand, if health care institutions and professionals, and organ retrieval organizations, seek to recover organs over the express objections of the family, litigation may well ensue.
45. Allan S. Brett, "Limitations of Listing Specific Medical Interventions in Advance Directives," *Journal of the American Medical Association* (1991), 266: 825-827; Joanne Lynn, "Why I Don't Have a Living Will," *Law, Medicine & Health Care* (1991), 19: 101-104.
46. Renate Justin, "The Values History: A Necessary Family Document," *Theoretical Medicine* (1987), 8: 275-281; Joan M. Gibson, "Reflecting On Values," *Ohio State Law Journal* (1990), 51: 451-471; Ben A. Rich, "The Values History: A New Standard of Care," *Emory Law Journal* (1991), 40: 1109-1181.
47. Wanzer, et al., note 38, p. 845.

48. Studies indicate that even when physicians do engage patients in discussions about advance directives, they infrequently deal with the values and attitudes that inform a patient's preferences for medical treatment. *See* James A. Tulsky, et al., "Opening the Black Box: How Do Physicians Communicate About Advance Directives?" *Annals of Internal Medicine* (1998), 129: 441-449.
49. Joan M. Gibson, note 46, pp. 455-471.
50. The advances of modern, high tech medicine have contributed significantly to America's dubious distinction as the most death-denying culture in human history. Ernest Becker, *The Denial of Death* (New York: The Free Press, 1973).
51. The "Five Wishes" document was created by a private, non-profit organization called Aging with Dignity. It is based in Florida and maintains a website at www.agingwithdignity.org.
52. Even physicians who are accustomed to caring for critically ill and dying patients have acknowledged how emotionally wrenching can be the decision and the process of withdrawing life support. Miles J. Edwards, Susan W. Tolle, "Disconnecting a Ventilator at the Request of a Patient Who Knows He Will Die: A Doctor's Anguish," *Annals of Internal Medicine* (1992), 117: 254-256.
53. Some of the SUPPORT investigators, after further analysis of the data, have arrived at the conclusion that the fundamental assumption underlying SUPPORT was fallacious. That assumption was that "the course of care for the seriously ill hospitalized patient is the result of individual, patient-level decision-making that could be improved with better counseling and information." The fallacy of that assumption purportedly lies in the recognition that the course of care, at least in hospital ICU's, is "shaped largely by how the care system is organized and by the interpersonal meanings ascribed to various cues and signals that shape the predictable patterns of care." Joanne Lynn, *et al.*, "Rethinking Fundamental Assumptions: SUPPORT's Implications For Future Reform," *Journal of the American Geriatrics Society* (2000), 48: S214-221.
54. For the most thorough discussion and analysis of prognostication in the care of seriously ill patients *see* Nicholas Christakis, *Death Foretold: Prophecy and Prognosis in Medical Care* (Chicago: University of Chicago Press, 2000).
55. SUPPORT investigators have arrived at the following dismal conclusion: "In these seriously ill patients, [advance directives] did not substantially enhance physician-patient communication or decision-making about resuscitation. This lack of effect was not altered by the PSDA or by the enhanced efforts in SUPPORT." Current practice patterns indicate that increasing the frequency of [advance directives] is unlikely to be a substantial element in improving the care of seriously ill patients. Future work to improve decision-making should focus upon improving the current pattern of practice through better communication and more comprehensive advance care planning. Joan Teno, *et al.*, "Advance Directives For Seriously Ill Hospitalized Patients: Effectiveness With the Patient Self-Determination Act and the SUPPORT Intervention," *Journal of the American Geriatrics Society* (1997), 45: 500-507.

Chapter 6

CONSTITUTIONAL LIBERTY AND PRIVACY: THE SUPREME COURT AND THE PHYSICIAN-PATIENT RELATIONSHIP

1. INTRODUCTION

In this chapter and the one that follows, the close interrelationship between law and biomedical ethics as a branch of applied philosophy can be discerned in a number of fundamental respects. There has been a tendency to characterize the Supreme Court decisions that will be our focus in these chapters – *Griswold v. Connecticut, Roe v. Wade*, and the other abortion cases prior to *Planned Parenthood v. Casey* – as having to do primarily with the family. However, *Griswold* and a subsequent case, *Eisenstadt v. Baird*[1] upholding the right of unmarried persons to access contraceptive information and devices, were fundamentally about the liberty of the individual, in consultation with a physician, to decide whether or not to conceive or bear a child. Often such decisions are made within the context of marriage or an intimate relationship with another person, but not always or necessarily. We should begin, however, with at least a provisional explication of the concept of privacy within the domain of constitutional jurisprudence.

2. LIBERTY AND PRIVACY IN CONSTITUTIONAL DISCOURSE

In both philosophy and constitutional law, the concepts of liberty and privacy can best be captured in the image of the double helix, for while one can carefully distinguish them at various points along a continuum, a complete understanding of how they operate in society at any given moment depends upon an acknowledgment of their inextricable intertwinings. Liberty, as a preeminent American value, predates the Constitution, having been enshrined in the stirring words of *The Declaration of Independence*: "We hold these truths to be self-evident, that all men are created equal, that they are endowed by their Creator with certain unalienable Rights, that among these are Life, Liberty, and the pursuit of Happiness." Thereafter, the preamble to the Constitution of the United States includes among its primary objectives "[to] secure the Blessings of Liberty." It is in the amendments to the Constitution, however, where we find a fuller expression of liberty and begin to discern its kindred spiritship with privacy in the American political psyche. For example, the Fourth Amendment, part of the original Bill of Rights, recognizes the right of the people to be secure "in their persons, houses, papers, and effects, against unreasonable searches and seizures," requiring warrants based upon probable cause. In the progression from the preamble to the Fourth Amendment the Constitution reveals that privacy (freedom from unreasonable searches and seizures of one's person, property and effects) is understood to be an essential aspect of the "blessings of liberty."

When we come to the Fourteenth Amendment, one of those which followed the Civil War, we find references to "the privileges or immunities of citizens of the United States," and a further assurance that persons may not be deprived of "life, liberty, or property, without due process of law." As we shall see, it was in the Fourteenth Amendment that the seeds of Constitutional privacy were sown, but would not sprout for decades to come.

The canonical expression of liberty in an opinion of the United States Supreme Court appears in a dissenting opinion by Justice Brandeis in *Olmstead v. United States*, wherein he states in words worth quoting in their entirety:

> The makers of our Constitution undertook to secure conditions favorable to the pursuit of happiness. They recognized the significance of man's spiritual nature, of his feelings and of his intellect. They knew that only a part of the pain, pleasure, and satisfaction of life can be found in material things. They sought to protect Americans in their beliefs, their thoughts, their emotions and their sensations. They conferred, *as against the government, the right to be let alone* – the most comprehensive of rights and the right most valued by civilized men.[2]

Interestingly, *Olmstead* was a case involving governmental wiretapping. Therefore, what Brandeis was most concerned about in the case before him, and which explains which words in the quoted passage he chose to emphasize, was governmental intrusion into personal communications made with an expectation of privacy. Nevertheless, "the right to be let alone" also connotes respect for individual liberty or autonomy – freedom from unwarranted interference from others in the pursuit of personal happiness. By suggesting the subtle and nonmaterial aspects of the lives of persons that are within the realm of Constitutional privacy, Brandeis was, perhaps, articulating a respect for liberty and privacy grounded on the concept of personhood. As we shall find when we consider some of the most important passages in the privacy cases of the last 35 years, a concept of the nature of persons in a democratic society seems to be at the heart of the Court's concerns. Respect for personhood offers a means of recognizing and explaining, rather than ignoring or obliterating, the dyadic relationship between liberty and privacy in the lives of persons.[3] The positing of a personhood paradigm for understanding certain aspects of Constitutional jurisprudence, particularly those relating to cases involving core issues in biomedical ethics, requires that we consider some of the philosophical treatment of personhood.

3. THE PHILOSOPHICAL DIMENSIONS OF PERSONHOOD

Even a brief review of the philosophical literature on the nature of persons would be beyond the scope of this work. However, because we are primarily concerned with the implications of personhood for understanding Constitutional liberty and privacy, there are certain perspectives on persons that are particularly relevant to our inquiry. Indeed, one of the preeminent philosophical analyses of persons is the product of the same individual whose political philosophy most profoundly influenced the framers of the Constitution, i.e., John Locke.

"Person," Locke famously observed, "is a forensic term."[4] What Locke means by this observation is not self-evident. One interpretation is that a "person" is as much an ascriptive as a descriptive term. That is, persons are not natural "kinds" that we identify and describe, but rather, a status that we ascribe to beings who possess certain attributes. Another interpretation, not necessarily incompatible with the previous one, is that the most significant implication of the ascription of personhood to someone is that they have particular moral worth and a concomitant moral accountability. Persons are beings who can be properly praised and blamed according to the nature of their acts and omissions. Implicit in the notion of moral accountability is individual autonomy or liberty, since one cannot be held morally accountable for conduct over which one does not have any control.[5]

Persons in the Lockean sense are not only responsible for what they do, but also for who they are. When Brandeis posited an intent of the framers of the Constitution to protect Americans in their beliefs, thoughts, emotions and sensations, he imputed to them a singularly Lockean sense of personhood. Such protections are essential to the flourishing of Lockean persons because it is through the exercise of these faculties that persons shape their own identities.

Indeed, it is through the formulation and pursuit of fundamental life plans and projects that persons in the Lockean sense both cultivate and express their unique identities. This perspective on the nature of persons has been aptly described by one prominent contemporary philosopher as Locke's "punctual self."[6] The term refers to the capacity for reflection by the self as subject on the self as object. The philosopher, Charles Taylor, suggests that the punctual self's capacity to engage in this radical form of disengagement and self-objectification is integral to the forensic nature of Lockean personhood. As we shall see, the self-defining and self-actualizing activities of the punctual self are a special concern of the privacy cases decided by the Supreme Court.

4. PERSONHOOD AND SUBSTANTIVE DUE PROCESS

The concept of Constitutional substantive due process was introduced in Chapter 2. The substantive nature of due process came into its full flowering when the Court struggled to articulate, as the spare phrases of the Constitution could not, what Constitutional law scholar Laurence Tribe describes as protected rights of personhood, and more particularly, human being, that are "not merely contemplated but required."[7] In order to understand the import of Tribe's assertion, one must re-introduce the concept of natural law and natural rights. The reason is that the substantive dimension of due process draws heavily upon natural rights – those that persons possess by virtue of their essential nature and not by virtue of their citizenship. So strong was the belief of those who drafted and ratified the Constitution in the existence of these inalienable rights that some perceived no need to specify them in the body of the document. However, as the United States in the Twentieth Century became vastly more complex and diverse, it also became alienated from the political and philosophical traditions of those who brought the Constitution into being. This alienation forced the Court to struggle anew with unenumerated yet fundamental rights of persons in a free society.

5. GRISWOLD V. CONNECTICUT: CONTRACEPTION AND PRIVACY

In the *Griswold* case, the Executive Director and one of the medical directors of the Planned Parenthood League of Connecticut challenged the constitutionality of a Connecticut statute which made it a crime to use a contraceptive device or to counsel or assist another person to do so. No distinction was made by the statute between married and unmarried persons.

Both of the challengers had been found guilty and fined as accessories, and their convictions upheld by the Supreme Court of Connecticut.[8]

5.1 The Opinion of the Court

In writing the Opinion of the Court, Justice Douglas noted that "This law . . . operates directly on an intimate relation of husband and wife and their physician's role in one aspect of that relation."[9] The case thus raised issues of privacy within the context of both the marital relationship and the physician-patient relationship.

The Court cast a wide net in an effort to ground its holding – that the Connecticut statute was unconstitutional – on prior decisions interpreting the First and Fourteenth Amendments. Two decisions in particular were considered to be of significance. In the first, *Meyer v. Nebraska*[10] the Court had struck down a state statute prohibiting the teaching of foreign languages in elementary school. In the second, *Pierce v. Society of Sisters*,[11] the Court held that the state could not require that all children attend public school. Both of these opinions were written by the same Supreme Court justice (McReynolds), and both drew upon unenumerated rights implicit in the concept of liberty:

> Without doubt . . . liberty denotes not merely freedom from bodily restraint but also the right of the individual to contract, to engage in any of the common occupations of life, to acquire useful knowledge, to marry, establish a home and bring up children, to worship God according to the dictates of his own conscience, and generally to enjoy those privileges long recognized at common law as essential to the orderly pursuit of happiness by free men.[12]

This passage can be successfully mined for much more than an unenumerated constitutional right to educate one's children in ways that may be disfavored by the state. Implicit in the right to marry and bring up children is the right to marry and not bring up children, a right which can be materially compromised if access to effective birth control counseling and devices is foreclosed by the state. Moreover, invocation of "those privileges long recognized at common law" appears to be a reference to the "privileges and immunities" clause of the Fourteenth Amendment, which provides that "No State shall make or enforce any law which shall abridge the privileges or immunities of citizens of the United States."

Justice Douglas relies on *Meyer, Pierce*, and a number of other cases as support for the proposition that "specific guarantees in the Bill of Rights have penumbras, formed by emanations from those guarantees that help give them life and substance. Various guarantees create zones of privacy." After citing the First, Third, Fourth, Fifth, and Ninth Amendments, and cases decided under them, Justice Douglas maintains: "These cases bear witness that the right of privacy which presses for recognition here is a legitimate one. The present case . . . concerns a relationship lying within the zone of privacy created by several fundamental constitutional guarantees."[13] That relationship, of course, is the marital one. Not content with a newly recognized right of privacy, in the concluding paragraph of the Opinion of the Court, Justice Douglas asserts that "[w]e deal with a right of privacy older than the Bill of Rights . . . Marriage is a coming together for better or worse, hopefully enduring, and intimate to the degree of being sacred."[14] Stated in this way, Justice Douglas appears to be saying that such a right of privacy is inherent in the natural rights of persons, rights which transcend government. So conceived, privacy is a right which governments must recognize but cannot confer upon persons, at least if they are to be respected in accordance with their nature and status according to natural law theory. As we shall see in subsequent Supreme Court decisions raising issues of privacy, there are radically divergent viewpoints among members of the Court with regard to the respect for persons that the state must show.

5.2 The Goldberg Concurrence

In a concurring opinion written by Justice Goldberg and joined by Chief Justice Warren and Justice Brennan, the emphasis is placed on the Ninth Amendment:

> To hold that a right so basic and fundamental and so deep-rooted in our society as the right of privacy in marriage may be infringed because that right is not guaranteed in so many words by the first eight amendments to the Constitution is to ignore the Ninth Amendment and to give it no effect whatsoever.[15]

Later in his concurrence, Justice Goldberg goes on at great length to embed the concept of marital privacy in the "private realm of family life which the state cannot enter."[16] Such characterizations are problematic for a number of reasons. They suggest that married persons have privacy interests greater than unmarried persons. Thus, it required a separate Supreme Court decision to affirm the right of unmarried persons to have access to contraceptive devices.[17] Furthermore, they suggest that the Supreme Court's privacy jurisprudence does not reach beyond the procreational decisions confronting the family, such as when, how, and with whom to raise children. Thus, we shall see that when the Court confronts Nancy Cruzan's right to decline artificial nutrition and hydration through her parents as proxy decision makers, the concept of Constitutional privacy is abandoned and Fourteenth Amendment liberty interests are invoked. Finally, they fail to confront the ultimate question of whether Constitutional liberty and privacy are two separate concepts. The Supreme Court has been subjected to extensive legal as well as philosophical critique for its failure to provide such a conceptual analysis.[18]

This invocation of the Ninth Amendment as the source for a right to privacy that includes procreational autonomy is controversial, largely but not exclusively because of the tenuous status of the Ninth Amendment in Constitutional history and jurisprudence. Simply stated, conservative judges and Constitutional theorists harbor serious concerns about judicial review based upon unenumerated rights. Nonetheless, the language of the Ninth Amendment presents one of the most direct and unequivocal references to the concept of natural rights to which many of the framers of the Constitution subscribed. Consider the following articulation:

> In a classical liberal theory of rights, rights define a sphere of moral jurisdiction that persons have over certain resources in the world – including their bodies. This jurisdiction establishes boundaries within which persons are free to do what they wish . . . Given this conception of rights – a conception in keeping with that held at the time of the framing of the Ninth Amendment – it is simply impossible to specify in advance all the rights we have . . . Instead of authorizing a search for particular rights, the Ninth Amendment can be viewed as establishing a general constitutional presumption in favor of individual liberty.[19]

On this view, Constitutional privacy delineates the sphere of moral jurisdiction within which persons, as natural law theory conceives of them, are sovereign. Within that sphere, their exercise of individual liberty (autonomy) is constrained only by their own moral sentiments and practical judgment. Society at large, or members thereof, may question a person's grounds for decision and reasons for acting, but government may not interfere.

5.3 The Implications of Griswold

Griswold has come to be regarded as the first foray by the Supreme Court in the modern (post World War II) era into the realm of Constitutional privacy. In striking down what even the dissenting justices agreed was an anachronistic and "uncommonly silly" state law, the Court laid the foundation for what would soon come to be the much more well-known and perennially controversial recognition in *Roe v. Wade* of the right to an abortion.

Constitutional conservatives or strict constructionists have been able to call into question the Supreme Court's privacy jurisprudence because of *Griswold's* reliance upon unenumerated rights conjured out of "penumbras, formed by emanations" from the Bill of Rights. Such unfortunate phrasing suggests that judicial legerdemain is at work, the conjuring of rights out of smoke and mirrors that would, in the words of Justice Steward, "have caused James Madison no little wonder."[20]

Perhaps because most people outside of the Catholic Church considered contraception, especially among married persons, to be an uncontroversial matter, the *Griswold* decision did not initially evoke strong criticism, even from conservative commentators.[21] For some, such as Robert Bork, there was a delayed reaction. At first he declared that "the idea of deriving new rights from old is valid and valuable," and only years later decided that it is "an unprincipled decision, both in the way it derives a new constitutional right and in the way it defines that right, or rather fails to define it."[22] Not until Constitutional privacy embraced abortion were the full implications of *Griswold* realized.

ENDNOTES AND REFERENCES

1. 405 U.S. 438 (1972).
2. 277 U.S. 438, 478 (1928) (Brandeis, J., dissenting).
3. It must be acknowledged at the outset that the Court has not openly embraced a right of personhood that incorporates liberty and privacy, prompting one commentator to observe that "it does not distinguish between privacy and autonomy and may be treating them both as aspects of 'the right to be let alone.'" Louis Henkin, "Privacy and Autonomy," *Columbia Law Review* (1974), 74: 1410-1433.
4. John Locke, *An Essay Concerning Human Understanding*, Peter H. Nidditch, ed., (Oxford: Cambridge University Press, 1975), Book II, Chapter 27, § 26.
5. For an important philosophical discussion of this point *see* Harry G. Frankfurt, "Freedom of the Will and the Concept of a Person," *Journal of Philosophy* (1971), 68: 5-20.
6. Charles Taylor, *Sources of the Self* (Cambridge: Harvard University Press, 1989), pp. 159-176.
7. Laurence H. Tribe, *American Constitutional Law*, 2nd ed., (Mineola, N.Y: Foundation Press, 1988), p. 1308. Laurence Tribe's Chapter on "Rights of Privacy and Personhood" is an outstanding resource worthy of careful study.
8. For a detailed historical account of the events leading up to the litigation in Griswold see David J. Garrow, *Liberty & Sexuality* (New York: MacMillan Publishing Co., 1994), Ch. 4.
9. Griswold v. Connecticut, 381 U.S. 479 (1965).
10. 262 U.S. 390 (1923).
11. 268 U.S. 510 (1925).
12. 262 U.S. 390, 399 (1923).
13. 381 U.S. 479, 486-487 (1965).
14. 381 U.S. 479, 486-487 (1965).
15. 381 U.S. 479, 491, (1965) Goldberg, J., concurring.
16. 381 U.S. 479, 495, (1965) citing Prince v. Massachusetts, 321 U.S. 158, 166 (1944).

17. Eisenstadt v. Baird, 405 U.S. 438 (1972) held that the right to be free from governmental intrusion into "matters so fundamentally affecting a person as the decision whether to bear or beget a child is an *individual* privacy right."
18. *See, e.g.,* Henkin, note 3.
19. Randy E. Barnett, "Reconceiving the Ninth Amendment," *Cornell L. Rev.* (1988), 74: 1-42, at 34-35.
20. 381 U.S. 479, 530, (1965) Steward, J., dissenting. For a contrary assessment *see* Barnett, note 19.
21. Even the Catholic journal *Commonweal* observed that "The entire round of court struggles was unnecessary, a dubious tribute to the power of a determined minority to impose their moral values on others." *Commonweal,* June 25, 1965, pp. 427-428.
22. This dramatic change in attitude is pointed out in David J. Garrow, *Liberty and Sexuality* (New York: MacMillan Publishing Co., 1994).

Chapter 7

THE DANCE OF INTIMACY: ABORTION, MEDICAL ETHICS AND THE CONSTITUTION

1. INTRODUCTION

Abortion is a relatively simple medical procedure with profound and controversial moral and legal implications. These attributes are reflected in the Supreme Court decision, *Roe v. Wade,* that galvanized the attention of American society when it was issued in 1973, and has continued to be a flash point and a call to arms for anti-abortion and pro-choice interest groups ever since. Neither the significance of the decision nor its place in American constitutional law can be adequately understood or fully appreciated outside of its historical context. For that reason, before beginning our consideration of that pivotal case, its anomalous historical background and the ambiguous, some might say ambivalent, perspective on abortion that it reflects, should be explored.

2. ABORTION AND THE SANCTITY OF LIFE – THE LESSONS OF HISTORICAL EXEGESIS

Many of the contemporary opponents of abortion are religious and political conservatives who ground their opposition on the sanctity of human life. Interestingly, however, in the *Old Testament,* there is no suggestion that a fetus has even so much as any protectable interests, let alone a right to life. The Ten Commandments in the book of *Exodus* declare that "Thou shall not commit murder," whereas other books in the *Old Testament* impose the penalty of death on a woman who conceives out of wedlock. Thus, the ending of the life of a fetus was clearly not deemed to constitute murder. In the Gospels of the *New Testament*, the silence on the subject of abortion is equally deafening.[1]

The argument that abortion constituted murder, or the wrongful taking of a human life, developed in the early and medieval Catholic Church. Thomas Aquinas suggested that at some time between 40 and 90 days of gestation the fetus became "formed" and hence the consequences of an abortion became much more significant. Even then, however, abortion of a fetus was distinguishable from the murder of a person born alive. The casuists, for example, prioritized the life of the pregnant woman over that of her fetus. Even contemporary Catholic opponents of abortion refer only to such biblical resources as the injunction to love your neighbor as yourself.[2] Such reliance erroneously presupposes that at the time such scripture was written, a fetus was deemed to have the status of a neighbor, i.e., a fully developed fellow human being.

Ambivalence toward abortion also characterizes the medical profession. The "Oath of Hippocrates" calls upon the physician not only to eschew surgery, sexual relations with patients, and the breaching of patient confidentiality, but also euthanasia and abortion.[3] As previously noted, and contrary to common assumptions, the authenticity of the Oath as the work of Hippocrates or any of his followers has been called into serious question. A theory that has gained such contemporary currency that it is even noted by Justice Blackmun in the Opinion of the Court in *Roe v. Wade* is that only the disciples of the Pythagorean school of philosophers actually adhered to its tenets.[4] Many Greek physicians at the time of Hippocrates (roughly 460-370 B.C) performed acts of abortion without incurring legal, ethical, or professional sanctions.

The role of the law in shaping the contours of the morality of abortion is most graphically illustrated in the United States between 1867-1973.[5] Until the middle of the Nineteenth Century, abortion prior to quickening – the time when the first movements of the fetus are detected, usually in the fourth or fifth month of pregnancy – was common. The focus on quickening can be traced through the English common law to earlier philosophical and theological arguments, particularly those of Thomas Aquinas, about when human life begins.[6] A clear consensus prevailed in the Western world, including the Catholic Church, by the early modern period, that prior to quickening abortion did not constitute the wrongful taking of a human life.

Despite the starkly moral and religious rhetoric that has pervaded the abortion debate in the last 30 years in the United States, the trend toward criminalization of abortion in the United States that began after the Civil War had little to do with any profound shift in public opinion about the ontological status of the fetus. Instead, the crusade against abortion, led by the medical profession, reflected issues of professionalism, the role of women in society, and, perhaps most remarkably, immigration in American society. The medical profession, which had not been nationally organized until the creation of the American Medical Association (AMA) in the mid-Nineteenth Century, saw the legal regulation of abortion as a means of marginalizing midwives, homeopaths, and the purveyors of "home remedies" as well as commercial preparations for inducing an abortion. Since the medical profession in this era was predominantly male, it is not surprising that the developing organized opposition to abortion had a decidedly anti-feminist cast. The male chauvinism that fanned the flames of organized medicine's crusade for the enactment of abortion laws had become open and notorious by the time of the publication of an 1871 report of the AMA Committee on Criminal Abortion, which characterized any woman who sought an abortion as "unmindful of the course marked out for her by Providence." The same report sought to recruit husbands into organized medicine's newly declared war against abortion by warning: "Let not the husband of such a wife [who would seek an abortion] flatter himself that he possesses her affection. Nor can she in turn ever merit even the respect of a virtuous husband. She sinks into old age like a withered tree, stripped of foliage; with the stain of blood upon her soul, she dies without the hand of affection to smooth her pillow."[7] Clearly, if the prevailing argument against abortion at that time were that it constituted the taking of innocent human life, the purple prose of the AMA would have read quite differently.

Interestingly, there were also anti-Catholic and anti-immigration strains to the anti-abortion crusade. An often-quoted tract by the crusading anti-abortion physician Horatio Storer warned that inasmuch as abortion was "infinitely more frequent among Protestant women than among Catholic," the question confronting us is whether the West (i.e., America) will "be filled by our own children or by those of aliens? This is a question which our women must answer; upon their loins depends the future destiny of the nation."[8] Similarly, Protestant religious opposition to abortion reflected concerns about the declining birthrates in its congregations rather than the moral or theological implications of the procedure.

3. THE MEDICALIZATION AND CRIMINALIZATION OF ABORTION

As contradictory as it might first appear, the enactment of antiabortion laws resulted in both medicalization and criminalization of the procedure. As a matter of strict statutory declaration and as a general rule, both pre- and post-quickening abortions were illegal. The general rule, however, was tempered with an important exception for when it was deemed necessary, in the opinion of a physician, to preserve the life of the woman. The presence of these laws in most states, declaring as a matter of public policy much more than as a matter of public morality, that abortion was illegal except when a physician deemed it necessary to preserve the life of the pregnant woman, appeared to serve two purposes. First, the number of abortions performed would be significantly reduced, and second, any abortions performed would be by licensed physicians based upon a determination of medical necessity.

As a matter of historical fact, the result was quite different. Remarkably, the number of abortions performed in the United States in the late Nineteenth and early Twentieth Centuries did not change. According to one source, as many as one out of every three pregnancies was terminated by an induced abortion (2 million each year), or a rate seven to eight times as high as today.[9] In a strict legal sense, very few of these abortions were performed in compliance with either the letter or the spirit of the typical anti-abortion statute. Most were performed by physicians for women whose social and economic status afforded them access to the best available medical care and were facilitated by an extremely loose definition of what was life-threatening to the pregnant woman. Poor women and those in rural areas without access to a sympathetic physician went to nonphysician abortionists or self-induced the abortion. Both avenues often led to fatal results.

Given the significant number of abortions being performed, only a small percentage of which could genuinely be deemed necessary to preserve the life of the pregnant woman, it is remarkable how few prosecutions (and even fewer convictions) there were under the antiabortion laws.[10] Sporadically there would be police raids on abortion clinics. However, the absence of any sustained public outcry over the continuing availability of abortions, particularly among women of means, strongly suggests a high tolerance for the political hypocrisy of using laws as moral window dressing. In this regard, anti-abortion laws in the century prior to *Roe v. Wade* were very much like anti-contraception laws in the century before *Griswold*. They existed to serve certain political ends, but not to be enforced consistently as though society was in earnest about eradicating the practices which they purported to proscribe.

4. SETTING THE STAGE FOR ROE V. WADE

Advances in medical science, particularly in the post World War II period, made it increasingly more difficult for physicians to finesse the "medical necessity" of many abortions sought by their pregnant patients. The number of pregnancies that actually posed a significant risk of death or serious physical harm to the pregnant woman had been greatly reduced over the last eighty years. With the rise of the modern hospital and the increase of procedures performed there, it became much more difficult for the abortion decision to be solely between a woman and her physician. Hospitals created review boards and charged them with the responsibility of determining, on a case by case basis, whether an abortion was necessary. Physicians became more and more uncomfortable with the ambiguity in most anti-abortion laws about what constituted a life-threatening situation that would justify terminating a pregnancy. They began to openly acknowledge as a profession the need to revisit the legal status of abortion.

In 1959, the American Law Institute, a highly regarded group of legal scholars and attorneys, proposed a revision to its Model Penal Code, which served as a model for a wide variety of criminal statutes. The revision provided that an abortion could be performed if, in the opinion of two physicians, one of three situations existed: (1) a substantial risk that continuation of the pregnancy would gravely impair the physical or mental health of the mother, (2) a grave physical or mental defect in the fetus, or (3) the pregnancy had been the result of rape or incest.[11] These revisions actually reflected the circumstances under which many physicians believed that it was appropriate to perform a therapeutic abortion, as documented by a study of 24 California hospitals that appeared the same year as the Model Penal Code proposal.[12] The momentum to make abortion law clearly and unambiguously consistent with current medical knowledge and practice was growing.

Often in American society, stark and compelling individual cases galvanize public sentiment in ways that thoughtful and persuasive argument does not. In 1962, such a case came to the forefront of the public consciousness in the plight of Sherri Finkbine, a 29-year-old pregnant mother of four. When she learned from a newspaper article that her fetus was at risk of severe defects because she had taken the drug Thalidomide, she consulted her physician, who confirmed her fears and strongly recommended an abortion. Because of the documented cases of severe birth defects in England, where the drug had been approved and where her husband had in fact acquired it for her, the physician anticipated a prompt approval by the hospital committee, the members of which he had informally polled about her case.

What transpired thereafter provides clear and convincing evidence of the problems inherent in the legal and medical status quo. The Finkbines sought to alert other pregnant women who might be at risk through their local newspaper, the *Arizona Republic*. When her story became public, the hospital administrator interceded and canceled the procedure. The stated reason was that he had been unable to secure sufficient assurances from the local prosecutor that legal action would not be forthcoming. The prosecutor, in turn, blamed the court because his request for a review of the matter was turned away on the grounds that no genuine case or controversy had been presented. The Finkbines then traveled to Sweden in order to obtain an abortion "in a more favorable legal climate." When the procedure was performed, the fetus turned out to have been severely deformed.

The Finkbine case, and an outbreak of German measles between 1962 and 1965 that resulted in many children with severe birth defects, significantly raised public awareness of the fact that abortions were not merely sought by promiscuous women who were unwilling to accept the responsibilities of motherhood. Increasingly, the law was seen as a real impediment to the medical profession's ability to respond to the legitimate medical needs of some pregnant women. However, as the advocacy for reform of abortion laws became more organized and vocal, so too did that of the abortion opposition, which in the 1960's consisted primarily of the Catholic Church. Breaking the relative silence that had characterized the abortion question between roughly 1880 and 1950, the Vatican publicly castigated Sherry Finkbine in the media both during and after her pursuit of an abortion in Sweden because "the victim was a human being." The Catholic Church's growing militancy in defense of existing abortion laws in the United States reintroduced the issue that had lain dormant for so long – the ontological status of the fetus, particularly prior to viability or even to quickening. The Vatican now asserted that human life begins at conception, so that no preference could be given to the life of the pregnant woman at any stage of pregnancy. By the early 1970's, when the movement for the repeal or major reform of the antiabortion laws had gained widespread and respectable support, the National Right to Life Committee was founded by a Cincinnati physician. Joining forces with the Catholic Church, the two groups fought against the growing public support for abortion with lurid and graphic images of aborted fetuses, a tactic that continues to characterize the so-called pro-life movement.

5. ROE V. WADE

We begin with a review of the facts and the holding of *Roe*, and then proceed to a discussion of its place in medical and constitutional jurisprudence. "Jane Roe" (Norma McCorvey) was a resident of Texas at the time she sought to terminate her pregnancy. At that time, the law in Texas was essentially as it had been for more than 100 years, i.e., abortion was illegal unless a physician certified that it was necessary to preserve the life of the pregnant woman. Roe initiated her legal action in federal court against Henry Wade, the district attorney of Dallas County, seeking a declaration that the Texas criminal abortion statute was unconstitutional on its face[13] and an injunction prohibiting him from pursuing its enforcement. Roe argued that under the First, Fourth, Fifth, Ninth, and Fourteenth Amendments she had a constitutional right to procure an abortion by a "competent, licensed physician, under safe, clinical conditions.[14]

The Opinion of the Court, written by Justice Blackmun, after disposing of procedural considerations such as justiciability and standing, moves to a detailed historical analysis of abortion in Western civilization. Among his conclusions is that "at common law, at the time of the adoption of our Constitution, and throughout the major portion of the 19th century . . . a woman enjoyed a substantially broader right to terminate a pregnancy than she does in most States today."[15] Justice Blackmun considered three factors which may have led to the constraints imposed on abortion in the United States in the previous 100 years: Victorian moral views about illicit sexual conduct, risks to the pregnant woman posed by abortion, and the purported interest of the state in protecting fetal life. "It is with these interests, and the weight to be attached to them," Blackmun asserts, "that this case is concerned."[16] The decision in *Roe* quite clearly stakes out a compromise position in what has been aptly characterized as a "clash of absolutes."[17] On one side is the argument of the plaintiff-appellant and the authors of several *amicus* briefs that a woman's right to obtain an abortion is absolute and unqualified at any stage of pregnancy, for any reason or for no reason she should be required to articulate. On the other side is the extreme right-to-life view that because human life begins at conception, the states can and should categorically prohibit abortion, at least to the extent of the Texas statute challenged in *Roe*.

The Opinion of the Court does not choose from among the various amendments to the Constitution a single one upon which to ground "a woman's right to choose." Rather, it simply states that from whatever portions of the Constitution the right to privacy is derived, it surely "is broad enough to encompass a woman's decision whether or not to terminate her pregnancy."[18] At least in part, this conclusion is bolstered by the Court's determination that the word "person" as used in the Fourteenth Amendment does not include the unborn.[19] However, despite the legal fact that a fetus is not a person within the meaning of the Constitution, the state's interest in protecting fetal life may become sufficiently compelling at some point prior to birth so as to justify laws regulating abortion. The critical question then becomes when during the course of pregnancy do the interests of the state become so compelling that it may regulate the abortion procedure without violating the Constitutional right of the pregnant woman to decide whether or not to terminate her pregnancy.

The mechanism the Court adopts in order to provide a fulcrum upon which to balance these competing interests is the trimester system. The only permissible restriction upon an abortion that a state may impose during the first trimester is that it be performed by a licensed physician. The only permissible restrictions during the second trimester are those reasonably related to concerns about the woman's health. In the third trimester, when the fetus is deemed to be viable (capable of surviving outside the womb), the state may regulate abortion almost as extensively as had Texas prior to *Roe, i.e.,* prohibiting it altogether except when it is determined by a physician to be necessary in order to preserve the life or health of the mother. Prior to the presumed viability of the fetus in the third trimester, the abortion decision is exclusively that of the woman, in consultation with her physician.[20]

Roe v. Wade is probably the most controversial decision of the Supreme Court in the Twentieth Century, and one of the most criticized by legal academics since the decision in *Lochner v. New York.*[21] The basic lay argument against *Roe* contests the proposition that a fetus is not a person with a right to life protected by the Constitution, or that personhood aside, human life begins at conception and it is human life rather than personhood that the Constitution protects.[22] The general thrust of the legal critique of *Roe* is one of constitutional interpretation. The contention is that there is no language in the Constitution from which one can reasonably derive a right to abortion, particularly one meriting the status of fundamental, so that states must demonstrate a compelling interest in order to infringe upon it. To assert this argument is, of course, to implicitly challenge the *Griswold* decision (see Chapter 6), for the decision in *Roe* draws heavily from its reasoning and analysis. But the critics of *Roe* must also challenge the long line of substantive due process cases mentioned earlier which recognized the significance of "unenumerated Constitutional rights" and the power they have to curtail state legislation that would infringe upon them. Thus, the strict constructionist approach would not only invalidate the right to abortion recognized in *Roe*, but also the right to contraceptive information and devices recognized in *Griswold v. Connecticut*[23], the right to interracial marriage recognized in *Loving v. Virginia*[24], and the right to exercise control over the education of one's children recognized in *Pierce v. Society of Sisters*[25] and *Meyer v. Nebraska*[26]. None of these decisions can point to specific language in the Constitution to support their ultimate conclusions.

In the nearly twenty years between 1973 and 1992, the year of the last major Supreme Court decision on abortion, the Court decided a plethora of cases brought to challenge the efforts of states to test and potentially extend the limits on legislation that *Roe* had imposed. With conservative Republicans in the White House from 1980 to 1992, abortion opponents went from cautiously optimistic to quite hopeful that new appointments to the Court would ultimately produce a five-vote majority in favor of over-ruling *Roe* and returning the issue of abortion to the more tractable state legislatures. When the case of *Webster v. Reproductive Health Service*[27] was decided in 1989, there seemed to be a very real prospect that the end of *Roe* was near, given the extent to which it was the target of critical commentary by a majority of the sitting justices.

In *Webster*, the Court considered three elements of a Missouri abortion statute: (1) a restriction on the performance of abortions in public institutions, even when they would be paid for with private funds, (2) a preamble declaring that "the life of each human being begins at conception," and (3) a requirement that tests for fetal viability be performed when the woman seeking an abortion is believed to be 20 weeks into her pregnancy. In upholding the constitutionality of the Missouri statute, the Court demonstrated how deeply divided the justices were. In order to arrive at a plurality opinion, the statute was construed so as to leave to the discretion of the physician the decision of whether or not tests to determine fetal viability were indicated. As for the declaration about when human life begins, the plurality finessed the hard question by finding that this language was nothing more than a permissible statement of a public policy that favored childbirth over abortion. There now appeared to be four clear votes to overrule *Roe*, four votes to uphold *Roe*, and one vote, that of Justice O'Connor, which would determine the future of abortion law in the United States. Then came the appointment and confirmation of the ultra-conservative justice Clarence Thomas, and a few months thereafter, a decision by the Court to review the case of *Planned Parenthood of Southeastern Pennsylvania v. Casey*.[28]

6. ABORTION, PERSONHOOD, AND SUBSTANTIVE DUE PROCESS

The *Casey* decision, the essence of which is a plurality opinion authored by Justices O'Connor, Kennedy, and Souter, purports to affirm "*Roe's* essential holding." That holding, according to the plurality, has three elements. The first is a woman's right to choose to have an abortion prior to viability and to obtain one without undue interference from the state. The second is the recognition of the state's power to restrict abortions after fetal viability, so long as an exception is provided for pregnancies endangering the woman's life or health. Third, the affirmation that from the beginning of the pregnancy the state has legitimate interests in protecting both the health of the woman and the life of the fetus that may become a child. It is the third element that constitutes, in the opinion of some commentators, the most significant departure from or reinterpretation of *Roe*. This is because *Roe* had been interpreted as disputing the contention that the state had any legitimate interest in the fetus prior to viability that would justify encroachment on a woman's right to terminate a pregnancy.

The woman's right to choose is grounded by the plurality opinion in the liberty interest protected by the substantive dimension of the Fourteenth Amendment due process clause. Noting that abortion remains one of the most divisive of moral and political issues, the opinion goes on to state that the Court's obligation "is to define the liberty of all, not to mandate our own moral code." Thus, the fundamental constitutional issue was viewed as whether the State can resolve what are essentially "philosophic questions" so as to deprive a woman of any choice in the matter. The Opinion of the Court then provides a startling disquisition on the inter-relationship between personhood and constitutional liberty:

> Our law affords constitutional protection to personal decisions relating to marriage, procreation, contraception, family relationships, child rearing, and education. Our cases recognize 'the right of the individual,

> married or single, to be free from unwarranted governmental intrusion into matters so fundamentally affecting a person as the decision whether to bear or beget a child.' Our precedents 'have respected the private realm of family life which the state cannot enter.' These matters, involving the most intimate and personal choices a person may make in a lifetime, choices central to personal identity and autonomy, are central to the liberty protected by the Fourteenth Amendment. At the heart of liberty is the right to define one's own concept of existence, of meaning, of the universe, and of the mystery of human life. Beliefs about these matters could not define the attributes of personhood were they formed under the compulsion of the State.[29]

Language with such sweeping breadth and deep philosophical import was not necessary for the Court to review and rule upon the challenged provisions of the Pennsylvania abortion statute. Perhaps it was deemed essential, however, to a ruling which reaffirmed a woman's right to choose and sought thereby to discourage further attempts by the anti-abortion interest groups and the United States government through the Solicitor General to persuade the Court to overrule *Roe.*

While the constitutional foundation for this paragraph is the liberty interest protected by the Fourteenth Amendment, rather than the right of privacy recognized in *Griswold* and applied in *Roe*, it conveys intimations of what Laurence Tribe has referred to as "the animating paradox of the right of privacy: it is revered by those who live within society as a means of repudiating the claims that society would make of them. It is a right that has meaning only within the social environment from which it would provide some degree of escape."[30] Throughout the history of the United States, states have undertaken to impose constraints upon fundamental life decisions by individuals, such as what professions they may enter, whom they may marry, whether or not they may decline to procreate, and as we shall consider in a subsequent chapter, whether they must undergo medical treatment which they do not wish to receive. It is with regard to such fundamental life choices that constitutional liberty and privacy stand as a bulwark against state action that unreasonably invades the physical, psychological, and personal integrity of its citizens to a degree that compromises their dignity and autonomy as persons in a free society.

What to many had appeared to be an integral element of the decision in *Roe*, the trimester framework, was jettisoned in *Casey* as not part of its essential holding. The problem with that framework, according to the plurality opinion, is that it undervalues (indeed, completely denies) the state's interest in fetal life during the first trimester. According to *Casey*, the state should be able to regulate abortion throughout pregnancy in ways that reinforce respect for fetal life, so long as they do not create an "undue burden" on the woman's right to choose. Thus, in the language of the Court:

> What is at stake is the woman's right to make the ultimate decision, not a right to be insulated from all others in doing so. Regulations which do no more than create a structural mechanism by which the State, or the parent or guardian of a minor, may express profound respect for the life of the unborn are permitted, if they are not a substantial obstacle to the woman's exercise of the right to choose.[31]

This new perspective on the nature and extent of the woman's right to choose empowers the state to undertake noncoercive stratagems of moral suasion designed to persuade a pregnant woman to continue the pregnancy, even in situations in which doing so may be against her individual best interests. Indeed, the Court specifically repudiates two prior decisions[32] to the extent that they would preclude a state from requiring physicians to convey to women seeking abortions information about the fetus and the impact of the abortion upon it.

It is significant that in no other instance does the state attempt to dictate to physicians the content of the information that they must convey about any particular procedure or treatment in the process of obtaining the patient's consent. But the implications of the Court's position on this issue are clear – it would certainly be within the purview of the state to dictate to physicians what information they must disclose about any or all procedures. The Court offers as an example that there is no impediment to the state requiring that for there to be an informed consent to a kidney transplant the recipient must be provided with information about the risks of the procedure to the donor as well as to him or herself.

The remainder of the *Casey* decision applied the undue burden test to the other provisions of the Pennsylvania law, upholding a mandatory 24-hour waiting period, a parental notification requirement with a judicial bypass provision, and a physician reporting requirement, while declaring unconstitutional a spousal notification requirement. Thus, the court moved from the lofty language about personhood and the right to define one's own concept of the meaning of human life to a pragmatic analysis that purported to justify state anti-abortion propaganda and waiting periods as not unduly burdensome. We shall find in Chapter 8 that the promise of the personhood disquisition quoted above failed to be realized when the Court at last confronted the ultimate instance of the purported right to die – physician assisted suicide.

Justice Blackmun understandably believed that *Casey* had severely wounded the woman's right to choose that *Roe* had recognized. Yet it remains to be seen what constitutes the greatest threat to this right – conservative jurists who are sympathetic to the opponents of abortion, or the medical profession itself, that has withdrawn in wholesale fashion not only from the provision of abortion services but also from the responsibility to teach the next generation of physicians how to safely and effectively perform them.[33]

7. CONCLUDING REMARKS

The implications of fetal personhood, or a fetal right to life, transcend the abortion cases. They implicate as well the extent to which the state is empowered to intervene in the lives of pregnant women who have not elected to have an abortion, but whose personal behavior is deemed to create an unreasonable risk of harm to the life or health of the fetus. One of the stronger feminist critiques of contemporary medicine has related to what feminists believe to be the false "maternal-fetal conflict."[34] They perceive it largely as a political ploy or power play on the part of the male-dominated medical specialty of obstetrics and gynecology to reign in, when necessary with judicial assistance, pregnant women who fail to follow "doctors' orders." These "orders" can range from a failure to refrain from the use of tobacco, alcohol, or drugs to the refusal to submit to prenatal fetal surgery or a caesarian section. The remedies courts have been urged to fashion, and some have indeed pursued, include mandatory injunctions,[35] actions by child protective services,[36] civil commitment,[37] and criminal prosecution.[38] Given feminist ethics' close association with an ethics of care, it should come as no surprise that these authoritarian and coercive approaches to high risk situations in pregnancy are uniformly resisted by feminist bioethics.[39]

The cases dealing with purported maternal-fetal conflicts do not reflect sophisticated and carefully nuanced bioethical analysis. Particularly those involving pregnancies in the third trimester tend to focus on the viability of the fetus and then propose to balance the risk of harm to the fetus posed by the pregnant woman's behavior with the infringement on the liberty and privacy interests of the woman that would be required in order to eliminate or minimize the risk of fetal harm. Slippery slope concerns are unavoidable, inasmuch as many lifestyle choices pose some (albeit limited) risks when they are undertaken by pregnant women. It is not difficult for the advocates of pregnant women to hypothecate totalitarian scenarios in which all pregnant women are, at some point in pregnancy, taken into a form of protective custody in which every aspect of their lives is dictated by a regimen prescribed by a panel of distinguished obstetricians.[40]

Thus far, the courts have failed to arrive at any consensus as to how to balance such interests as the pregnant woman and her fetus may have that are genuinely placed in conflict by her lifestyle choices. The analysis is complicated not only by the perennial disputes over the extent to which a fetus at any stage of development has interests, but also by the argument of some officials that the state has its own interest in fetal health. The state's interest is said to be particularly strong in those instances in which the pregnant woman both intends to give birth and to continue to engage in such practices as the use of substances known to produce harm to a fetus that will seriously affect the quality of the life of the child to be.[41]

We shall find in the next chapter's consideration of state interests relating to patient refusals of life-sustaining medical interventions that the courts, including the Supreme Court, have been loathe to engage in any critical analysis of the interests the states have asserted in the lives of their citizens. Whether this is primarily a failing of the courts or of the advocates who appear before them is difficult to definitively determine, but there are some clues which we will consider at the appropriate time. There are other instances in medical jurisprudence in which the courts have demonstrably failed to even make an effort to engage with the larger philosophical issues involved in the cases before them. One example is an early case involving surrogate motherhood, in which a state supreme court insisted on trying to wedge a square peg into a round hole by analyzing the case and reaching a decision as though it were a garden variety matter of custody or adoption.[42] Another example is from the many cases that have refused to recognize the wrongful life claims of children born with severe genetic impairment.[43] We now turn to those cases in which courts have, with great reluctance, confronted the ultimate questions raised by the fundamental fact of human mortality, and the war currently being waged against it with such vigor by the medical profession.

ENDNOTES AND REFERENCES

1. Greg E. Pence, *Classic Cases in Medical Ethics*, 2nd ed. (New York: McGraw-Hill, Inc., 1995), pp. 146-147.
2. John T. Noonan, Jr. "An Almost Absolute Value in History," in John T. Noonan, Jr., ed.., *The Morality of Abortion: Legal and Historical Perspectives* (Cambridge, MA: Harvard University Press, 1970).
3. Ludwig Edelstein, *The Hippocratic Oath: Text, Translation, and Interpretation* (Baltimore: John Hopkins Press, 1943).
4. Ben A. Rich, "Postmodern Medicine: Deconstructing the Hippocratic Oath," *University of Colorado Law Review* (1993); 65: 77-136.
5. A detailed historical account of abortion law and practice during this period of American history has recently been provided in Leslie J. Reagan, *When Abortion Was A Crime: Women, Medicine, and Law in the United States, 1867-1973* (Berkeley: University of California Press, 2000).
6. Stephen J. Heaney, "Aquinas and the Presence of the Human Rational Soul in the Early Embryo," *The Thomist* (1992), 56: 19.

7. Quoted in Brief of 281 American Historians as Amici Curiae Supporting Appellees in *Webster v. Reproductive Health Services*, 109 S. Ct. 3040 (1989), at pp. 17-18.
8. Horatio Storer, "The Criminality and Physical Evils of Forced Abortions," *Transactions of the American Medical Association* (1865) at p. 16.
9. Reagan, note 5.
10. It has been reported that between 1911 and 1930 in Minnesota there were 100 indictments for abortion and 31 convictions. In Michigan between 1893 and 1932 there were 156 indictments and 40 convictions. Laurence H. Tribe, *Abortion: The Clash of Absolutes* (N.Y.: W.W. Norton & Co., 1990), p. 35.
11. David J. Garrow, *Liberty and Sexuality* (N.Y.: MacMillan Publishing Co., 1994) at p. 277.
12. Herbert L. Racker, Ralph J. Gampell, "Therapeutic Abortion: A Problem in Law and Medicine," *Stanford Law Review* (1959), 11: 417-455.
13. There are two types of constitutional challenge to a statute – facial and as applied. To challenge a statute as unconstitutional "on its face" is the more formidable of the two, since it requires the challenger to demonstrate that there is no conceivable instance in which the statute could be applied without violating the constitution in some way. An "as applied" challenge need only show that when the statute is enforced upon the challenger, some constitutional provision is violated.
14. At least by implication, Roe was also asserting that this right was unqualified by the fact that she was single, not at risk of serious harm from carrying the fetus to term, and financially unable to travel to another jurisdiction where an abortion under such circumstances would be available to her. Roe v. Wade, 410 U.S. 113, 120 (1973).
15. 410 U.S. 113, 140 (1973).
16. 410 U.S. 113, 152 (1973).
17. The phrase is the title of Laurence Tribe's important work on the subject of abortion. *See* note 10.
18. 410 U.S. 113, 153.
19. It is interesting to note that while the Court confidently concludes that those who drafted and ratified the Constitution did not intend to include the unborn when they used the word "person," it also denies that it is in any sense resolving the question of when human life begins, a question about which the Court notes there remains a lack of consensus. The Court goes on to state that while the unborn are not persons in the Constitutional sense, human life may well begin at some point before birth. The key ontological question that the Court does not confront directly is the moral and legal status of the fetus prior to becoming a person but after human life is acknowledged to have begun.
20. At one point in the Opinion of the Court Justice Blackmun writes: "The decision vindicates the right of the physician to administer medical treatment according to his professional judgment up to the points where important state interests provide compelling justifications for intervention. Up to those points, the abortion decision in all its aspects is inherently, and primarily, a medical decision, and basic responsibility for it must rest with the physician." 410 U.S. 113, 165-166. The woman's right to choose seems to take a back seat to professional autonomy. Such a characterization of what is really at issue in *Roe* overlooks a fundamental fact about abortion that we noted at the beginning of this Chapter – it is a medical procedure that in most instances is performed for nonmedical reasons relating to the woman's determination that she does not wish to bear a child. Subsequent decisions by the Supreme Court authored by other justices do not contain such physician-oriented language.
21. 198 U.S. 45 (1905). The *Lochner* decision has come to characterize both an era, "the *Lochner* era" (roughly 1897-1937), and a judicial process, "Lochnerizing." The most salient feature of *Lochner* and similar cases was the propensity of the Supreme Court to strike down state legislation regulating labor by applying strict judicial scrutiny of the relationship between the state's goals and the statutory means by which they were pursued.

22. Ronald Dworkin provides an extended and carefully reasoned argument that opposition to abortion, even in its most strident and uncompromising forms, is not based upon a firm and settled conviction that a fetus is a person with an unqualified right to life. Rather, the opposition to abortion, as well as the recognition by those who are pro-choice of the moral significance of a decision to undergo an abortion, are based upon a widespread belief that a fetus embodies a form of human life that is in some quasi-secular sense "sacred." Ronald Dworkin, *Life's Dominion* (New York: Alfred A. Knopf, 1993).
23. 381 U.S. 479 (1965).
24. 388 U.S. 1 (1967).
25. 268 U.S. 510 (1925).
26. 262 U.S. 390 (1923).
27. 492 U.S. 490 (1989).
28. 505 U.S. 833 (1992).
29. 505 U.S. 833, 851 (1992).
30. Laurence H. Tribe, *American Constitutional Law*, Second Edition (Mineola, N.Y.: Foundation Press, 1988), p. 1302.
31. 505 U.S. 833, 877 (1992).
32. Akron v. Akron Center for Reproductive Health, 462 U.S. 416 (1983); Thornberg v. American College of Obstetricians and Gynecologists, 467 U.S. 747 (1986).
33. More than 80 percent of the counties in the United States have no physicians who will perform an abortion, while some states have only a few physicians who do so. Only 12 percent of medical schools teach students how to perform a first trimester abortion during their obstetrics and gynecology training. Leslie J. Reagan, *When Abortion Was A Crime: Women, Medicine and Law in the United States* (Berkeley: University of California Press, 2000).
34. Dawn Johnson, "The Creation of Fetal Rights: Conflicts with Women's Constitutional Rights to Liberty, Privacy, and Equal Protection," *Yale Law Journal,* 1986; 95: 599; Bonnie Steinbock, "Maternal-Fetal Conflict and In Utero Fetal Therapy," *Albany Law Review*, 1994; 57: 781.
35. Jefferson v. Griffin Spalding County Hospital Authority, 274 S.E. 2d 457 (Ga. 1981).
36. In re Fetus Brown, 689 N.E. 2d 397 (Ill. App. 1997).
37. Minn. Stat. Ann. §§ 626.5561, 253B.05.
38. People v. Hardy, 469 N.W. 2d 50 (Mich. App. 1991).
39. *See generally* Joan C. Callahan and James W. Knight, "Women, Fetuses, Medicine, and the Law," in Helen Bequaert Holmes and Laura M. Purdy, eds., *Feminist Perspectives in Medical Ethics*, (Bloomington: Indiana University Press, 1992), pp. 224-239.
40. Efforts by employers to promulgate "fetal protection" policies that prohibited not merely pregnant but all fertile women from holding certain positions was held by the Supreme Court to violate Title VII of the Civil Rights Act of 1964 and the Pregnancy Discrimination Act in the case of Automobile Workers v. Johnson Controls, 499 U.S. 187 (1991).
41. A relatively comprehensive list of the behaviors that can cause fetal harm is provided in Reinesto v. Superior Court, 894 P.2d 733, 736-737 (Ariz. App. 1995).
42. In the Matter of Baby M, 537 A.2d 1227 (N.J. 1988).
43. Language which has been quoted in a number of these cases reads as follows: "[w]hether it is better never to have been born at all than to be born with even gross deficiencies is a mystery more properly to be left to the philosophers and theologians. Surely the law can assert no competence to resolve the issue, particularly in view of the very nearly uniform high value which the law and mankind has placed on human life, rather than its absence." Becker v. Schwartz, 386 N.E.2d 807 (N.Y. 1987).

Chapter 8

DEATH, DYING, AND THE RESPONSIBILITY OF THE ETHICAL PHYSICIAN

1. INTRODUCTION

Historically, we have assumed that the death of any living thing is a fact we discover, not a status that we confer. Physicians have been the traditional arbiters of death in that they have had the legal authority to pronounce or declare that someone is dead. But in most instances this was something of a formality. When someone ceased to breathe, cardiac arrest followed shortly thereafter, and within minutes the massive and irreversible death of brain tissue. This natural progression of events could not be halted or reversed by physicians until quite recently in the long history of medicine. The cessation of cardio-pulmonary function became a quite reliable test, ultimately to be performed by a licensed physician, in order to definitively determine that someone had died.

When we consider alternative ways to think about human death, terminology becomes crucial. We began to speak of a cardio-pulmonary *definition of* death, rather than *test for* death. This is problematic because there are four disparate yet related features about death which must be considered: a *concept* of death, a *definition* of death, a set of *criteria* for the determination of death, and *tests* to be performed in order to ascertain whether the *criteria* have been met. As we shall find, from the very first discussion of the need to re-think the process according to which human death is determined, semantic considerations have been pervasive and problematic.

Our concept of something is our general idea or understanding of it. As a matter of logic and consistency, our definition of death, our criteria for establishing that some formerly living entity is now dead, and our tests for confirming that the criteria have been met in a particular instance, should be derived from and consistent with our concept of death. It is possible, however, to articulate a strictly formal definition of death as a heuristic device with which to clarify our concept of death. Such a formal definition might read as follows: "Death is a complete change in the status of a living entity resulting in the irreversible loss of those characteristics that are essentially significant to it."[1]

In arriving at our concept of death, which should be the starting point, the task then becomes one of ascertaining those characteristics of a human being that are so essential to its nature that their permanent loss *should* constitute death. Regrettably, the most influential pronouncements on the subject of brain death do not proceed in this rigorous fashion, and as a result have been the cause of considerable confusion about how we think and speak of human death.

2. INTRODUCING BRAIN DEATH: THE AD HOC COMMITTEE OF THE HARVARD MEDICAL SCHOOL

The subject of brain death highlights a number of issues at the intersection of medicine, law and ethics. It was introduced by a seemingly innocuous report by an ad hoc committee of the Harvard Medical School ("Ad Hoc Committee") published in a 1968 issue of the *Journal of the American Medical Association*.[2] The thrust of the report was that advancements in medical science and technology had rendered the standard cardio-pulmonary definition of death – permanent cessation of spontaneous cardiac and respiratory function – inadequate. The first and perhaps most important advancement was mechanical ventilation, which allowed patients who would otherwise have died from respiratory failure to be maintained indefinitely, even though they had lost the capacity for spontaneous respiration. This life-sustaining technology has since been materially enhanced by an arsenal of new drugs to reverse a variety of cardio-pulmonary problems that previously would have led to cardio-pulmonary arrest and certain death.

The second development was that of organ transplantation techniques and technologies, including heart transplant with the advent of the bypass technology, that enabled artificial circulation of the patient's blood while the defective heart was removed and replaced with the donor's heart. If the donor heart could not be removed until the cardio-pulmonary criteria for death had been met, serious deterioration of the organ would take place, perhaps making the transplant ineffective. If, on the other hand, the patient might be declared dead by some other criterion, the organ or organs could be removed expeditiously, thereby maximizing the likelihood of a successful transplant.

The position taken by the Ad Hoc Committee was that when a patient was determined to be in an "irreversible coma," the patient could be declared dead, all life support withdrawn, and with the appropriate consent organs removed and made available for transplantation. "Irreversible coma" is one of a number of unfortunate uses of language in the report. For example, the authors of the report state that "our primary purpose is to *define* irreversible coma as a new *criterion* for death" (italics added). The report goes on to state: "responsible medical opinion is ready to adopt new *criteria* for pronouncing death to have occurred in an individual sustaining irreversible coma . . . If adopted . . . it can form the basis for change in the current legal *concept* of death" (italics added). Using terms like "definition" and "criterion" as if they were interchangeable is one of many reasons why brain death continues to this day to be problematic for physicians as well as lay persons. We need to carefully consider each of these elements independently and utilize appropriate terminology.

Because the Ad Hoc Committee report conflates "concept," "definition," and "criterion," one must look beyond mere terminology in order to ascertain whether there is a particular concept of death underlying the recommendation for a particular brain death formulation. This task is especially relevant to the subsequent determination of whether to adopt a whole brain or a higher brain death formulation. In the first paragraph of the Ad Hoc Committee report, one of the two reasons offered for the recognition of brain death is the great burden posed by "patients who suffer permanent loss of intellect." Later, in the commentary stage of the report, the Ad Hoc Committee notes that "modern resuscitative and supportive measures . . . can now restore 'life' . . . when there is not the remotest possibility of an individual recovering consciousness following massive brain damage."

Two years after publication of the report, Henry Beecher, the chair of the Ad Hoc Committee, in a paper presented to the American Association for the Advancement of Science on the subject of brain death, elaborated on the concept of human death. He indicated that what was most distinctive about human beings and essential to their nature is "the individual's personality, his conscious life, his uniqueness, his capacity for remembering, judging, reasoning, acting, enjoying, worrying, and so on." All of these attributes, he concludes, can be captured by a single word, "consciousness."[3] Beecher's reflections on the distinctive characteristics of human beings are precisely what is necessary in the development of a concept of death. Each one of these statements militates toward the higher brain, or cognitive functions, rather than the lower brain or vegetative functions such as regulating pulse, respiration, blood pressure, body temperature and the like. Yet when the Ad Hoc Committee set out the criteria for determining whether brain death has taken place, the focus shifted from the higher brain to the lower brain functions, and more particularly, unresponsiveness to painful stimuli, no spontaneous respiration, and complete loss of reflexes.

The internal inconsistencies of the report continue with its discussion of the nature of the problem and the proposed solution. Although initially acknowledging that "moral, ethical, religious, and legal issues" are involved, the report goes on to note that the law has traditionally looked to the medical profession for the criteria by which to determine death. Hence, the report reasons, so long as the medical profession adopts a consensus position in support of the new brain death formulation, "no statutory change in the law should be necessary." Implicit in this assertion is an assumption that resort to the new criterion for declaring a patient dead when "irreversible coma" is present in no way alters our basic concept of death. In order to fully appreciate the fallacy of this assumption, we need to consider the defense of the whole brain formulation of death subsequently offered by the President's Commission. First, however, we will briefly consider the remarkably swift and uncontroversial adoption of the Ad Hoc Committee's recommendations into law and public policy.

3. BRAIN DEATH AND THE LAW

Between the publication of the Ad Hoc Committee report in 1968 and the report of the President's Commission in 1981, some states amended their statutes to accommodate brain death (Kansas was the first in 1970) and several model acts were proposed. The most widely accepted of these has been the Uniform Determination of Death Act promulgated by the National Conference of Commissioners on Uniform State Laws in 1980. This version, adopted by 26 states, provides as follows:

> An individual who has sustained either (1) irreversible cessation of circulatory and respiratory functions, or (2) irreversible cessation of all functions of the entire brain, including the brain stem, is dead. A determination of death must be made in accordance with accepted medical standards.[4]

As the title of the act suggests, its purpose is to legally recognize the propriety of declaring a person dead when the clinical criteria support a finding that there has been a total and permanent cessation of brain function, as well as when there has been an irreversible cessation of circulatory and respiratory function. Such acts do not undertake to articulate a concept of death or to define death in a manner that is consistent with any presumed concept. However, one of the more thoughtful judicial decisions regarding the legal status of brain death explicitly noted the philosophical dimensions of the determination of death:

> With the ability of modern medical techniques to restore the function of vital organs or compensate for their nonfunction, medical decisions may be made based not on "scientific" fact but on the physician's *concept* of life and death. The decision by a physician as to whether a person is

> dead is thus not merely a medical, biological, or physical conclusion. It is, in part, a philosophical decision about what conditions *define* human life, combined with an empirical determination that those conditions are absent and not latent in a given case. The determination involves differentiating between human life and biological life, marking the dividing line between what constitutes human life and that which is purely mechanical . . . . (italics added).[5]

The Washington court held that it is for law to define the standard of death and for the medical profession to determine the applicable criteria, inasmuch as death is both a legal and a medical question. By declaring that brain death is an appropriate "standard of death," the court acknowledges making a judgment that "total unawareness and complete unresponsiveness are the equivalent of death addresses questions related more to philosophy and law and is not the exclusive domain of medicine." Interestingly, however, the court then promptly and without further analysis adopts the whole-brain death standard of complete cessation of all brain function, including the brain stem, without further critical inquiry into whether that standard for brain death (recommended by the Ad Hoc Committee) was most consistent with our concept of human death.

4. PRESIDENT'S COMMISSION REPORT – *DEFINING DEATH*

The President's Commission report considers at some length the various criteria that have been suggested as indicative of human death: cessation of the flow of vital fluids, the loss of bodily integration, and loss of the capacity for conscious experience. As for the first, the Commission concludes that "the flow of bodily fluids could conceivably be maintained by machines in the absence of almost all life processes: the result would be viewed by most as a perfused corpse, totally unresponsive to its environment."[6] More importantly, from the Commission's perspective, a patient who had experienced cessation of all brain functions could appear to be alive according to the cardio-pulmonary criteria so long as artificial life-support measures were continued. There is a crucial distinction, the Commission maintains, between the "perfused corpse" that is dead according to the whole brain formulation that it espouses, and the patient with a functioning lower brain but no neocortical function:

> The startling contrast between bodies lacking all brain functions and patients with intact brain stems (despite neocortical damage) [is that] the former lie with fixed pupils, motionless except for chest movements

> produced by their respirators [whereas] the latter cannot only breathe, metabolize, maintain temperature and blood pressure . . . on their own but also sigh, yawn, track light with their eyes, and react to pain or reflex stimulation.[7]

Having offered a distinction that it considers to be of sufficient significance to justify treating the former as a corpse and the latter as a patient, the Commission nevertheless concedes shortly thereafter that "it is not easy to discern precisely what it is about patients in this latter group that makes them alive while those in the other category are not." It is precisely this point that advocates of higher brain (also referred to as neocortical or cognitive) death emphasize. By seizing upon the role of the lower brain and brain stem in integrating the physiological functions of the human body, the Commission seems to suggest that this constitutes the essence of what it means to be alive as a human being.

By contrast, advocates of higher brain death argue that the sighs, yawns, and flinches that distinguish the permanently unconscious patient devoid of neocortical function from the perfused corpse that has undergone whole brain death are of no significance whatsoever because a person no longer resides in that body. The answer of such advocates to the question "Does anyone survive neocortical death?" is "no."[8] The most serious flaw in the Commission's analysis is that it fails to adequately focus on our conception of human death, and instead considers only criterial candidates. The suggestion that most people conceive of human death as the permanent loss of the capacity to auto-regulate the body's physiological processes is belied by the Ad Hoc Committee commentary regarding the loss of intellect and consciousness.

5. A BRIEF AND TENTATIVE CONCEPTUAL ANALYSIS OF BRAIN DEATH

One of the chief distinctions between the life of human beings and other forms of life is that it partakes of two dimensions – biological life and personal life. Before the advent of modern medical science and technology, these two dimensions of human life were entirely coextensive. Now, however, a patient who suffers the complete loss of higher brain function, and hence will never have another conscious experience, can be sustained through their natural life span in a persistent vegetative state (PVS). It is accurate to say in such instances that the patient's human personal life is over, because the capacity for conscious experience is a necessary condition for personhood. What remains in a kind of suspended animation is human biological life of the most rudimentary form – the janitorial functions of the body – artificially supported by medical technology.[9]

If we return to the strictly formal definition of death mentioned in the introduction, the President's Commission, and by implication all of the states through the adoption of brain death criteria consistent with the whole brain death formulation, have taken the position that the characteristic essentially significant to a human being is the capacity to auto-regulate non-cognitive bodily functions. Hence, only the permanent loss of the capacity to perform that function shall be deemed to constitute death.

Advocates of higher brain death maintain that such a conceptual analysis completely discounts the most essential attribute of human beings, their capacity for conscious experience. Discussions of death in the philosophical literature, as one might expect, undertake the kind of conceptual analysis that naturally follows from the formal definition of death. They explore those qualities that are essential to human beings as we understand their nature. The majority lean strongly toward the loss of the capacity for conscious experience rather than to loss of the capacity for auto-regulation of physiological functions as the defining feature of human death. The following are several examples:

> "To conceive of one's own death, it would be quite sufficient to conceive of the cessation of one's experiences."[10]

> "My death is simply the permanent cessation of my consciousness."[11]

> "Human death is conceived of as an 'experiential blank,' an infinitely long period without any consciousness."[12]

Such philosophers are not disputing the biological fact that individuals who have permanently lost the capacity for consciousness may nonetheless exhibit physiological functioning of a rudimentary nature, some of which may actually be auto-regulated by persisting lower brain and brainstem structures. What they do contest is that the persistence of these homeostatic physiological functions is how we conceive of life.

To argue persuasively for a higher brain conception of human death does not require that we currently possess the technological capacity to declare someone dead according to a definition that is consistent with such a conception. A definition of human death based upon the higher brain death (cognitive or neocortical) conception would be that death is the permanent loss of the capacity for conscious experience. Criteria for the determination of death so-conceived and so-defined would be the total destruction or complete dysfunction of the neocortical structures of the brain which sponsor conscious experience. The tests administered in order to ascertain whether the criteria had been met in any particular instance would be those which can, with a reasonably high degree of accuracy, measure the functioning and/or structural integrity of the neocortex.

While positron emission tomography (PET scan) may be a prime candidate for such a test, at this stage of its development there may still be an unacceptably high risk of false-positive determinations of death. However, simply because the current level of technology permits us to make reasonably accurate determinations of death based only upon the whole-brain formulation does not justify the prevailing efforts to re-conceptualize human death and to define it as the permanent loss of the body's capacity to auto-regulate basic physiological functioning.

A number of other maneuvers have been employed in an effort to demonstrate the unintelligibility, or at least the impracticality of the higher brain formulation of death. With regard to the former, a few critics have insistently argued that death is necessarily a biological concept, and therefore one simply cannot use death to denote anything other than the demise of physiological functioning. Since "person" does not necessarily refer to a particular biological organism, the death of the person (or the end of human personal life) is a purely figurative manner of speaking, like the "death of tragedy," and not a proper basis upon which to declare that a human organism has died. There are at least two plausible responses to this contention. First, because of the categorical distinctions (concept, definition, criteria, tests), it is not at all illogical to maintain that our concept and definition of human death may be predominantly non-biological, whereas the criteria and tests utilized in declaring human death may be biological. Furthermore, the higher brain formulation is firmly grounded on a recognition that the capacity for human personal life or for conscious experience is grounded on certain physiological functions – those of the neocortex. Even the whole brain formulation allows for the declaration of human death despite the fact that certain biological processes are still active in the human organism. The essence of the debate over the proper formulation of brain death (whole-brain versus higher brain) is first about whether human biological life or human personal life should most fully inform our concept and definition of death, and secondarily about which physiological functions are essential to that conception and definition.

A practical problem with the higher brain formulation is supposed to be that human bodies that continue to spontaneously maintain pulse, blood pressure and respiration by virtue of a functioning lower brain or brain stem cannot be declared dead. To do so would require that we be prepared to either bury (or presumably cremate) them while these physiological functions are underway, or wait indefinitely until they have ceased. This is a specious argument. Once a person has been declared dead, all manner of measures are undertaken to prepare the body for burial or other disposition which would be unacceptable prior to the declaration. Thus, once a person is declared dead on the higher-brain formulation, measures necessary to terminate residual physiological functions can be undertaken because we are no longer confronted with a patient but with a corpse. Those measures might constitute killing prior to the declaration of higher brain death, but certainly not subsequent to it.

Two states have enacted statutory exceptions to the declaration of death according to "neurological criteria" when the physician has reason to believe that to do so would violate the personal religious beliefs of the individual.[13] Such "conscience clauses" lend credence to the proposition that the determination of human death, at least since the advent of high-tech medicine, is not simply the ascertainment of a biological fact. Rather, as the President's Commission recognized, "the basic concept of death is fundamentally a philosophical matter" and "philosophical issues persist in the choice to define death in terms of organ systems, physiological functions, or recognizable human activities, capacities and conditions."[14]

The continuing ambivalence toward brain death is manifested in a number of ways. One example is the proclivity to describe clinical scenarios in which a patient is declared brain dead yet is thereafter "kept alive" (vital signs supported by medical technology) until some later date when the technology is withdrawn and the patient is allowed to die (cessation of vital signs). Such behaviors and characterizations, combined with the conscience clauses previously mentioned, raise an important ethical issue. When a physician definitively determines that a patient meets the criteria for brain death, does that fact create a duty to declare the patient dead, or merely an opportunity to do so? If it is the former, then what are we to make of the conscience clauses in these two statutes, and how, in other jurisdictions, should physicians accommodate the needs of families whose concept of death is inextricably linked to physiological functioning, regardless of whether it is spontaneous or supported by medical technology? If it is the latter, then can we really say that as a matter of law and social policy brain death is really death? We certainly do not encounter instances in which a physician has asserted any discretion not to pronounce a patient dead who had experienced the permanent cessation of pulse and respiration.

Depending upon what formulation of human death is adopted, certain decisions about when life-sustaining interventions should be discontinued are taken out of the area of discretion, inasmuch as no medical interventions are appropriate for a corpse. The whole-brain formulation leaves a much larger range of situations in which these often anguishing decisions must be made, in particular those involving patients such as Karen Ann Quinlan and Nancy Cruzan, who meet the higher brain but not the whole brain definition. Indeed, advocates of the currently prevailing whole brain definition insist that simply because their favored formulation does not permit a physician to declare a patient in a PVS to be dead in no way creates a legal or an ethical obligation to continue life-sustaining treatment, particularly when there is evidence that the patient would not wish to receive it. We turn now to that process of decision making.

6. WITHHOLDING AND WITHDRAWING LIFE-SUSTAINING TREATMENT

Judicial decisions about life-sustaining measures can be separated into two broad categories – patients with decisional capacity and patients without decisional capacity. I use the term 'decisional capacity' rather than the often-used term 'competence' because the two are distinguishable in an important sense. Whether or not a patient possesses decisional capacity is a determination which the law generally allows individual physicians to make. Competence, on the other hand, is a formal legal determination made by a court, which then results in the designation of a guardian or conservator of the incompetent ward. A patient need not be declared incompetent in order for a physician to look to others to serve as surrogate or proxy decision makers. Furthermore, most states now recognize, through statutory and/or case law, that resort to surrogate decision making does not require a judicial proceeding.

Given the history of informed consent in the United States as it moved from an innovative idea in 1957 to a well-recognized legal right in less than 20 years, it may come as a surprise that the period between the mid-1960's and the late 1980's was one in which even the rights of patients with decisional capacity to decline life-sustaining medical interventions were hotly contested. In large measure, as we shall see, these cases reflect the continuation of the struggle between patient autonomy and self-determination on the one hand, and the authority of the medical profession to pursue its traditional paternalistic practices on the other. In some instances, it is also difficult to determine with any reasonable degree of certainty whether we are dealing with a patient who does or does not possess decisional capacity. Part of the paternalistic bias of medicine, in which some judges became complicit, was that patients facing serious illness, particularly of an acute nature, almost invariably had their decision making capacity impaired.

An early case in point was *Georgetown College*[15] in which the university hospital sought emergency judicial authorization to transfuse a Jehovah's Witness patient over her objection and that of her husband. The college's attorneys were unsuccessful at the district court level. Rather than accepting the ruling and respecting the patient's right to refuse such treatment on religious grounds, the attorneys then petitioned a judge on the appellate court, the highly regarded J. Skelly Wright. Judge Wright went to the hospital to assess the situation for himself. His subsequently issued opinion and order allowing the transfusions to be made is replete with paternalistic shibboleths. In addition to presuming that a patient who has lost 60% of her blood volume cannot possess decisional capacity, Judge Wright determined that since the patient had a seven-month-old child, her refusal of life-saving treatment constituted abandonment, which it was within the *parens patriae* powers of the state to prohibit. He went on to suggest, without providing justification or analysis, that respecting her wishes would place the hospital and physicians at risk of legal liability. The exact opposite was quite clearly the case. Once the court ruled that a patient's refusal of treatment must be respected, there could be no viable claim against the hospital or physicians for failure to provide such treatment. Finally, and again without offering any authority, he opined that the patient's religious convictions merely precluded her (and presumably her husband as well) from consenting to a transfusion. In other words, he presumed without evidence that if she were transfused over her express objection, her religious position would not be compromised. Judge Wright's opinion and order in *Georgetown College* provides stark and compelling evidence of my prior contention that Judge Cardozo's statement in *Schloendorf* in 1914 that "every person of adult years and sound mind has the right determine what shall be done with his own body" had, in the years thereafter, died the death of a thousand legal qualifications.

A petition for rehearing was filed in the *Georgetown College* case, although it technically was moot since the transfusions had been performed pursuant to the prior order. The petition for rehearing was denied. However, several of the judges argued in separate dissenting opinions that rather than simply denying the petition for rehearing, the Court should instead dismiss the petition for lack of jurisdiction. A particularly impassioned opinion was written by Warren Burger, who later would become Chief Justice of the United States. Justice Burger essentially maintained that from the very beginning the Court should have dismissed the petition by Georgetown College on the grounds that there was no justiciable controversy. As he points out, "the touchstone to justiciability is injury to a legally protected right." In this case, there simply is no legally protected right of the hospital or its physicians to impose treatment on this patient against her will. Invoking the famous language of Justice Brandeis on "the right to be let alone," Burger then applies it to the facts of the case and concludes that such a right is not limited only to "sensible beliefs, valid thoughts, reasonable emotions, or well-founded sensations." It also includes "a great many foolish, unreasonable and even absurd ideas which do not conform, such as refusing medical treatment even at great risk."[16]

Gradually, over the next 25 years, as such cases continued to be litigated, a consensus developed among the courts of the United States in support of the right of adult patients to decline medical interventions necessary to save life. However, even in some of the more recent cases, dissenting judges invoke the language of Judge Wright to justify what they perceive as the state's compelling interest in preserving human life regardless of the contrary interests of the patient whose life is hanging in the balance.[17] When we discuss the *Cruzan* case subsequently, we will consider the bizarre manner in which the invocation of state interests has been used to separate a person's wishes with regard to his or her own life from the purported interest of the state in the sanctity of human life abstracted from the life of the person living it. The assertion of such countervailing state interests directly poses the rhetorical question that became the title of a Broadway play and subsequent motion picture – *Whose Life Is It, Anyway?*

One of the most compelling cases involving a patient with decisional capacity was that of Elizabeth Bouvia. At the time that her legal conflict approached its ultimate conclusion, Ms. Bouvia was a 28-year-old woman with a case of cerebral palsy so severe that she was quadriplegic and in constant pain from degenerative and crippling arthritis. She required constant care, and her last few years had amounted to a succession of unsuccessful searches for an adequate setting in which to receive that care. She ultimately came to a Los Angeles County public hospital, where the physicians and nurses responsible for her care determined that she was not ingesting adequate nutrition by mouth and hence required artificial nutrition and hydration. When Elizabeth Bouvia refused to consent, the attending physician threatened to initiate artificial nutrition and hydration against her will if necessary. Her refusal of this form of medical treatment was characterized as suicide, and acting (or omitting to act) in complicity with it was viewed by the health care professionals as assisted suicide.

Represented by *pro bono* legal counsel, Bouvia asked the probate court to issue an order prohibiting the provision of artificial nutrition and hydration over her express objection. Since no one involved in the care of Ms. Bouvia had ever suggested that she lacked decisional capacity or was afflicted by any mental condition that raised any doubt as to her competency, one might have expected that the strength of her legal arguments was overwhelming. After all, distinguished jurists by then had been opining for decades that "Anglo-American law starts from the premise of thorough-going self determination,"[18] and "every person of adult years and sound mind has the right to determine what shall be done with his own body."[19] Regrettably, however, as we have already seen in other cases, the core ethical value and legal principle underlying these ringing assertions – respect for the autonomy of the individual – continued to be compromised through an open and notorious collaboration of paternalistic physicians and patronizing judges. That was again the case when the probate court, while acknowledging that Elizabeth Bouvia was "rational," "sincere," and "fully competent," nevertheless allowed the force-feeding to proceed because of the negative impact her death would have on the hospital staff and the disabled population generally.[20]

Each of the four interests of the state that can be asserted in opposition to a patient's exercise of the right to refuse treatment appeared to be a factor in the judge's ruling. The origins of these interests are extremely obscure. Their initial iteration can be traced to a 1977 Massachusetts decision.[21] These interests, which have been subsequently adopted without justification or critical analysis, are listed as follows: (1) maintaining the sanctity of life; (2) preventing suicide; (3) protecting innocent third parties; and (4) upholding the ethical integrity of the medical profession. However, the Massachusetts court offered neither an explanation for the existence of these interests nor any authoritative source from which they were derived. As I have argued extensively elsewhere, the pedigree of and the rationale for these interests is extremely questionable.[22]

The state's interest in preserving life was certainly a factor, and so too was the interest in preventing suicide, at least so long as Bouvia's refusal of artificial nutrition and hydration was characterized as suicide. Since she had no children and her husband had left her, the probate court reached out to the group Advocates for the Developmentally Disabled, which had been holding candlelight vigils outside of the hospital, as an example of the innocent third parties whose interests would be harmed (or at least jeopardized) by allowing Bouvia to refuse artificial nutrition and hydration. Finally, the professionals responsible for her care were strenuously arguing that medical ethics required that an unnecessary death be avoided.[23]

Bouvia lost her appeal of the order, but subsequently was able to make other arrangements and left the hospital. Slightly over 18 months later, she was again in a public hospital in southern California and judged by her physicians to be receiving inadequate nutrition by mouth. Apparently assuming that the issue of force-feeding was forever *res judicata*, Bouvia was again subjected to tube feedings against her will. Her fate at the hands of another trial court judge was no different than before – he determined that her refusal of tube feedings was motivated not by a legitimate exercise of a constitutional right, but by an illegitimate desire to terminate her life.[24] On her appeal of this decision, a California appellate court at long last recognized that the right to refuse treatment was more than ringing phrases that were ultimately hollow at the core, and that it included refusal of treatment without which the patient might die.

The Court of Appeals did not refuse to consider the state interests as asserted by the public hospital and its medical and nursing staff, but considered them from a much different perspective than the previous courts, and one much more consistent with contemporary judicial thinking. Among the important points made by the court was that quality of life was a legitimate issue in the case – from the patient's perspective. It was not within the province of the state to declare what quality of life a patient must accept. The court also rejected the suggestion that Bouvia must continue to endure suffering she found intolerable in order to somehow reduce the risk that treatment might be withdrawn from some future disabled patient who might actually wish to receive it. The assertion of an unqualified state interest in preserving life, without any consideration of the pain and suffering entailed for the person who must live it, is not a correct interpretation of public policy.[25] The Court of Appeals also rejected the trial court's consideration of Bouvia's motives for refusing artificial nutrition, asserting: "If a right exists, it matters not what 'motivates' its exercise. We find nothing in the law to suggest the right to refuse medical treatment may be exercised only if the patient's *motives* meet someone else's approval."[26] This was precisely the point made by Justice Burger in his opinion regarding the *Georgetown College* case.

Right to life advocates have made much of the fact that after her victory in the appellate court, Elizabeth Bouvia did not in fact starve herself to death. The implication they would have us draw is that her caregivers knew her mind better than she did. A similar contention has been put forward by such groups, as well as by health care professionals who staff the burn centers of hospitals, with regard to the case of Dax Cowart. Cowart was a young man in his twenties who suffered severe burns over 67% of his body when a gas pipeline exploded. Over several years he persistently requested that the excruciatingly painful treatment be stopped and that he be allowed to die. In a complex story of medical and legal paternalism, Cowart continued to receive the medical treatment he refused and to be denied the day in court that he sought in order to establish that he possessed the decisional capacity to refuse treatment. A psychiatrist who conducted extensive discussions with him acknowledged that there was really no question Cowart had decisional capacity. Because Cowart ultimately left the hospital, married, obtained a law degree, and discontinued his unsuccessful efforts to end his life, he has been offered as living proof of the ultimate beneficence of such therapeutic belligerence. Dax Cowart continues to maintain otherwise.[27]

7. THE SUPREME COURT, INCOMPETENT PATIENTS, AND TREATMENT REFUSAL

As we have seen, respect for patient autonomy, both concurrent through informed consent and prospective through advance directives, has a confusing and conflicted history in American jurisprudence. The situation becomes even more complex and ambiguous when we confront cases of decisionally incapacitated patients without written directives. References to respect for their autonomy in health care decision making is admittedly hypothetical.

The case that the U.S. Supreme Court ultimately chose to consider was that of Nancy Cruzan, a woman in her mid-twenties who had been in a persistent vegetative state for nearly five years after sustaining severe traumatic injuries in a one-car accident. When her parents had come to terms with the fact that she would never regain consciousness, they determined, knowing her as they did, that she would not wish to have her physiological existence indefinitely prolonged by artificial nutrition and hydration. The Missouri State hospital where she was receiving treatment refused to discontinue artificial nutrition and hydration unless a court ordered them to do so. However, while the Cruzan family presented what a state trial court found to be clear and convincing evidence that Nancy would not wish to be kept alive in her present condition, it also directed the Missouri Attorney General to ask the state supreme court to review the case since it was one of first impression in Missouri.

A sharply divided Missouri Supreme Court took the highly controversial step of second-guessing the trial court's assessment of the evidence, holding that it did not meet the "clear and convincing" standard that was appropriate when a patient's life hangs in the balance. We have seen this disfavored disregard of the weight attached to evidence by a trial court before in the *Martin* decision by the Michigan Supreme Court. Furthermore, the Missouri court extensively reviewed the extent to which almost every other state had applied a best interests standard of review in those cases in which clear and convincing evidence was not available, only to reject that alternative by concluding that: (1) the constitutional right to privacy does not include "the provision of food and water to an incompetent patient," (2) because of the state's interest in the continuation of the patient's life regardless of its quality, the fact that Nancy has no hope of returning to a cognitive, sapient state is irrelevant, and (3) Missouri's interest in respecting the sanctity of Nancy's life outweighs her interest (as advocated by her surrogates) in having artificial nutrition and hydration discontinued.[28]

The Missouri Supreme Court decision in *Cruzan* is profoundly disturbing in many ways which cannot be appreciated by reading the *Cruzan* decision by the U.S. Supreme Court, which we will take up next. Consider, for instance, the following portion of a single paragraph of the majority opinion:

> But this is not a case in which we are asked to let someone die. Nancy is not dead. Nor is she terminally ill. This is a case in which we are asked to allow the medical profession to make Nancy die by starvation and dehydration.[29]

If Nancy Cruzan were dead, it would not be in anyone's power to let her die. Whether PVS, Nancy's condition, is a terminal illness or not is a medical judgment, not a legal one. Moreover, while a terminal condition was a condition precedent to following the dictates of a living will, Nancy did not have a living will. Her parents, as her legal guardians, had authority to consent to the initiation of artificial nutrition and hydration, and hence they had the authority to direct that it be discontinued. Withdrawing artificial nutrition and hydration pursuant to the instructions of a surrogate decision maker is not "making Nancy die."

At another point in the opinion of the court, the majority insists that "the state's interest is not in quality of life. The state's interest is an unqualified interest in life." Yet, having acknowledged that Nancy was in a PVS, which is a form of permanent unconsciousness, any discussion of her quality of life becomes meaningless. What the court is asserting is that the State of Missouri has an unqualified right to cause the biological life of a patient to be sustained as a means of demonstrating its vitalistic reverence for life.

The U.S. Supreme Court granted the Cruzan family's petition for review. A 5-justice majority held that although a Constitutional liberty interest in refusing unwanted medical treatment, including artificial nutrition and hydration, could be inferred from prior decisions of the Court, Missouri's interest in preserving human life (regardless of its quality) was sufficient to allow it to impose a clear and convincing evidence standard on surrogate decision making. The majority also found no constitutional infirmity in the Missouri Supreme Court's conclusion that the evidence presented by the Cruzan family at trial did not meet the clear and convincing evidence standard.

Strong dissenting opinions by Justices Brennan (joined by Justices Marshall and Blackmun) and Stevens challenged the legitimacy of the Court's determination that Missouri's imposition of a clear and convincing evidence standard in a case such as *Cruzan* violated no constitutional rights. Justice Brennan framed the issue in the following terms: "whether the Due Process clause allows Missouri to require a . . . patient in an irreversible persistent vegetative state to remain on life-support absent rigorously clear and convincing evidence that avoiding the treatment represents the patient's prior, express choice."[30] Arguing that the right of a patient to refuse medical treatment, a right that survives the loss of decisional incapacity, outweighs the interests of the state, individually or collectively, Justice Brennan challenged the proposition that the clear and convincing evidence standard imposed by Missouri was designed to assure that the treatment she received was consistent with her wishes. While there was some evidence in the record that Nancy Cruzan would not wish to have her life sustained in a PVS, evidence which the trial court found persuasive but the Missouri Supreme Court did not, there was absolutely no evidence in the record to support the argument that she would wish continuing treatment. Thus, it is reasonable to conclude that it is much more likely than not that such continued treatment was against her wishes. If all Missouri wished to do was to insure that the care the incompetent patient receives is likely to be consistent with her wishes, it would not completely discount all credible evidence that fails to meet an arbitrarily imposed and often impossibly high burden of proof.

What both of the dissenting opinions in *Cruzan* suggest is that Missouri's primary concern is not fidelity to Nancy Cruzan's wishes, but to its own abstract notions about the sanctity of all human life. Thus, Justice Brennan charges: ". . . Missouri and this Court . . . have discarded evidence of her will, ignored her values, and deprived her of the right to a decision as closely approximating her own choice as humanly possible. They have done so disingenuously in her name, and openly in Missouri's own."[31] Similarly, Justice Stevens argues: "This Court . . . permits the State's abstract, undifferentiated interest in the preservation of life to overwhelm the best interests of Nancy Beth Cruzan, interests which would, according to an undisputed finding, be served by allowing her guardians to exercise her constitutional right to discontinue medical treatment."[32]

The U.S. Supreme Court decision in *Cruzan* was but a major battle in a long war of attrition waged by the Cruzan family against the authority of the state to compel medical treatment they firmly believed she would reject. The family ultimately won that war. Because of the notoriety the case received, other colleagues and acquaintances of Nancy brought forward additional evidence that suggested she would not wish her life prolonged in a PVS. When that evidence was presented in another hearing before the trial court, it once again ordered that her feeding tube be removed. This time, the Missouri Attorney General declined to challenge the order, and five months following the Supreme Court's decision, Nancy Cruzan died.

At the beginning of the Twenty-First Century, the right of patients to decline life-sustaining treatment is well established, but as the cases we have considered reflect, that right is tenuous at best – in practice if not in theory – with regard to patients who lack decisional capacity. For the minority of persons who have executed written directives, particularly Durable Powers of Attorney for Health Care, there is now at least a reasonable prospect that care inconsistent with their values can be withheld or withdrawn. But for the vast majority who have avoided discussions of the subject, or those who have only spoken in general terms, their medical fate will be determined by a great many variables such as in what jurisdiction and by whom their care is provided.

The legacy of litigation over the right to refuse treatment, and particularly the authority of surrogate decisionmakers to assert that right on behalf of decisionally incapacitated patients, is in many ways a troubling one, for reasons that go well beyond the already mentioned refusal to critically analyze the "countervailing state interests." Consider, for example, the serendipitous manner in which such interests are asserted. A dispute must arise, between a health care institution or provider on one side and a patient or a patient's surrogate on the other, that ultimately leads to litigation. Furthermore, if the dispute is exclusively between private parties, then a representative of the state would have to be brought into the litigation. There is no proper way for state interests to be asserted by a court without completely sacrificing its role as an unbiased arbiter of the dispute. If states such as Missouri are so determined to demonstrate their reverence for the sanctity of all human life, why have they not instituted measures that would trigger their involvement in each and every attempt on the part of a patient or surrogate to prevent the administration of life-sustaining measures?[33] The cynical answer, of course, is that the elected public officials are unwilling to expend the significant political (not to mention economic) capital that would be required to insinuate the state into life and death decisionmaking across the board. They will simply accept what role the state can secure through the offices of non-elected appellate judges.

Another troubling aspect of these cases is the failure of the courts to place any burden of proof whatsoever on the states when asserting their interests or presuming to protect the interests of vulnerable patients. When, as was the case in *Cruzan*, the family made a *prima facie* case that Nancy would not have wished to be sustained in a PVS, reason and justice would seem to require that the burden of proof should shift to the State of Missouri to introduce competent, credible evidence that in fact she would have wished to be sustained indefinitely in a permanently unconscious state. Such evidence, of course, was nonexistent. Yet Missouri was fully prepared to impose continued existence upon her despite the fact that most adult Americans have indicated that they would not wish such treatment once they had permanently lost the capacity for conscious experience.

There is yet another extremely perplexing aspect to these right-to-die cases, which is why the states should have only interests that make it an adversary of its citizens. In a free and democratic society in which individual liberty and privacy are ostensibly sacred, states should have not merely an interest in, but a responsibility for preserving, protecting, and defending the right to decide, directly or through a duly designated surrogate, the nature of the medical treatment to which its citizens are subjected. Some courts, such as in *Bouvia*, have taken on this role. And Justices Brennan and Stephens, in their *Cruzan* dissents, expressed their views that it was monstrous for the State of Missouri to assert that it had an interest in keeping a patient like Nancy alive even if her wish would have been to die with dignity. But there has been precious little concern expressed about the perverse idea that under the Constitution of the United States the lives of citizens can legally and morally be held hostage to a vitalist state.

8. FROM TREATMENT WITHDRAWAL TO PHYSICIAN-ASSISTED SUICIDE – THE EUTHANASIA DEBATE

Physician-assisted suicide (PAS) is a form of what has generally been described as euthanasia or mercy killing. While the traditional view of euthanasia as a "good death" was one brought about for the best interests of the individual (from the individual's point of view), the term has been misappropriated for nefarious purposes, such as by Nazi Germany in the 1930's and 1940's.[34] Further complications arise from the delineation of at least four categories of euthanasia: voluntary active, voluntary passive, non-voluntary active, and non-voluntary passive. Withdrawing life-sustaining measures at the request of a patient with decisional capacity could be characterized as voluntary passive euthanasia, whereas withdrawing such measures from a patient without decisional capacity and without a clear advance directive could be characterized as nonvoluntary passive euthanasia. As we have seen, each of these have been deemed both ethical and legal under appropriate circumstances. Even those who oppose euthanasia as a matter of general moral or religious principle tend to accept the legitimacy of these practices. Voluntary active euthanasia would be the administration of a lethal injection pursuant to the request of a patient with decisional capacity. Whether the writing of a lethal prescription should be characterized as active or passive euthanasia is a debatable question. Those who seek to distinguish the two argue that the direct and immediate act causing the patient's death in the case of a lethal prescription is that of the patient, whereas in the case of a lethal injection it is that of the physician. Moreover, the lethal prescription is a necessary but not a sufficient condition for the patient's death, whereas the lethal injection is both. Those who see them as morally indistinguishable argue that "but for" the physician's action (injecting the lethal drug or writing the lethal prescription) the patient would not have died when and how she did.

Four factors have been pivotal in bringing the issue of physician-assisted suicide to the center of public attention and debate. Chronologically speaking, the first was the introduction of the practice of euthanasia (voluntary active) by physicians in the Netherlands, subject to a set of guidelines. The second was a series of assisted suicides by the former pathologist Jack Kevorkian in the State of Michigan. Several efforts to criminally prosecute Dr. Kevorkian resulted in jury verdicts for the defendant. He was ultimately convicted of second-degree murder in 1999, when he crossed the line from PAS to voluntary active euthanasia by administering a lethal injection to a patient with amyotrophic lateral sclerosis who wished to end his unbearable suffering. The third was a series of challenges to the constitutionality of state laws making assisted suicide a criminal offense, even when the "victim" is a terminally ill patient who seeks the assistance of a willing physician. Two of these cases found their way to the U.S. Supreme Court and will be the subject of the next section. The fourth was the enactment by referendum of the Oregon Death With Dignity Act, legalizing and regulating PAS. At least in the United States, the litigation of this issue has galvanized public debate of a bioethical issue to an extent not seen since *Roe v. Wade* did with regard to abortion.

9. THE SUPREME COURT AND PHYSICIAN-ASSISTED SUICIDE

In 1996, two federal circuit courts of appeal held that statutes making assisted suicide a crime were unconstitutional insofar as they precluded a physician from providing a patient with decisional capacity and in a terminal condition with a lethal prescription for the purpose of hastening their death. In *Compassion In Dying v. Washington*[35] the Ninth Circuit held that a Washington statute violated the Fourteenth Amendment liberty interest of such patients "in determining the time and manner one's own death." The Ninth Circuit took issue with the characterization of such cases as suicide, since the patient's death in the near future from the terminal condition was inevitable. The Second Circuit held that a similar statute in New York violated the Equal Protection clause of the Fourteenth Amendment because it treated terminal patients on life-support differently.[36] Such patients could determine the time and manner of their death by causing the life-support to be discontinued, whereas terminally ill patients who were not on life support could not obtain a lethal prescription. Because two influential federal circuit courts of appeal had overturned state laws making assisted suicide a criminal offense, and did so based upon two separate provisions of the Fourteenth Amendment, the Supreme Court was under considerable pressure to grant certiorari and review both decisions. What few commentators were able to predict was that a unanimous Court would reverse both of the lower courts.

Whereas the Ninth Circuit opinion recounted a history of suicide and assisted suicide reflecting ambivalence within and among societies, and a list of intellectual luminaries such as Seneca, Sir Thomas More, and David Hume who have supported suicide, the Supreme Court opinion written by Chief Justice Rehnquist noted the enduring disapprobation toward suicide and assisted suicide of the Anglo-American common law. The Supreme Court also took issue with the Ninth Circuit's framing of the issue, maintaining that the issue before it was "whether the 'liberty' specially protected by the Due Process Clause includes a right to commit suicide which itself includes a right to assistance in doing so."

When thus characterized, it becomes a foregone conclusion that such a right cannot be "deeply rooted in this Nation's history and tradition" or "implicit in the concept of ordered liberty," such that "neither liberty nor justice would exist if they were sacrificed."

The Supreme Court was equally unimpressed with the Second Circuit's application of the Fourteenth Amendment's Equal Protection Clause to physician-assisted suicide. The Court found the Second Circuit's analysis to be flawed because: "*Everyone*, regardless of physical condition, is entitled, if competent, to refuse treatment; *no one* is permitted to assist a suicide."[37] Thus, the Court shifted the focus from all competent, terminally ill patients seeking to hasten their death to all persons within the state.

These companion decisions by a unanimous Court and the concurrences that accompany them can be critiqued on a number of grounds. First, despite the fact that the Ninth Circuit found a Fourteenth Amendment liberty interest in some form of physician-assisted dying based upon an "as applied" challenge to the Washington statute, the Supreme Court, for its own unstated purposes, chose to treat the challenge as a facial one.[38] Through this disingenuous piece of judicial legerdemain, the Court transformed the burden of the physicians and terminal patients who initiated the challenge from the realm of the difficult to that of the virtually impossible. The ludicrousness of this maneuver by the Court becomes patently obvious in Justice Steven's concurrence, when he expounds at great length upon the sort of "as-applied" challenge that he would be likely to find meritorious. The type of facts he describes in the hypothetical case that might cause him to find a constitutional right to assisted suicide are precisely the facts presented by the named patient plaintiffs in *Glucksberg* and *Quill*. Moreover, these patients were not making a *facial* challenge to the Washington and New York laws as unconstitutional in all conceivable cases. Rather, they claimed only that *as applied* to their special circumstances – terminally ill patients with decisional capacity, free of clinical depression, enduring great suffering despite state-of-the-art palliative measures, and desirous of the assistance of a willing physician in the hastening of their proximate and inevitable deaths – these statutes violated their right to a humane, physician-assisted death.[39]

Justice O'Connor's concurrence contains the most glaring misstatement, or to be more charitable, misperception of reality. She blithely declares: ". . . in these States a patient who is suffering from a terminal illness and who is experiencing great pain has no legal barriers to obtaining medication, from qualified physicians, to alleviate that suffering, even to the point of causing unconsciousness and hastening death."[40] The documentation in the professional literature of the formidable barriers to effective pain management posed by the regulatory oversight of physician prescribing practices by state medical licensing boards and the federal Drug Enforcement Administration is, to put it modestly, overwhelming.[41] Perhaps realizing that his fellow jurist had seriously misrepresented the current state of affairs, Justice Breyer attempted to contain the problem by noting that: "We are also told that there are many instances in which patients do not receive the palliative care that, in principle, is available, but that is so for institutional reasons or inadequacies or obstacles . . . which do not include *a prohibitive set of laws*."[42] Later, Justice Breyer joins in Justice O'Connor's suggestion that if a future case came to the Court in which a law were shown to prevent the administration of drugs as needed to avoid pain at the end of life, then "the Court might have to revisit its conclusions in these cases." What remains unclear from the five concurring opinions that intimate support for what has since come to be characterized as "a Constitutional right to palliative care"[43] is how direct must be the barrier imposed by a law in order for it to raise legitimate Constitutional concerns.[44] While five of the nine justices on the Court appear to support assuring that dying patients receive as much medication as they require to be made comfortable, we know enough about their judicial philosophy to recognize that they will be extremely reluctant to inject the Court into problems that are not clearly caused or materially exacerbated by laws with constitutional infirmities. In Section 11 we will consider further the moral and legal status of aggressive palliative measures.

10. THE OREGON DEATH WITH DIGNITY ACT AND ITS OPPONENTS

The legislatures of a number of states have and will continue to grapple with proposed legislation to legalize and regulate physician-assisted suicide. No legislature has even come close to doing so, despite repeated public opinion polls showing that a slight majority of Americans believe that under some circumstances physicians should be able to provide assistance in hastening the death of a terminally-ill patient. The Oregon Death With Dignity Act, which regulates PAS under limited circumstances, became law through voter initiative, not through the usual legislative process that had received such praise in the concurring opinions of a number of Justices in *Glucksberg* and *Quill*.[45] Among the protections built into the Oregon law are: the patient must be over the age of eighteen and diagnosed with a terminal condition by two physicians, the patient must be referred to counseling if the physician believes the patient suffers from depression or a mental disorder, the patient must make two verbal and one written request for a lethal prescription with minimum periods separating the requests, the patient must administer the lethal prescription, and a reporting system must be in place for tracking those who receive PAS.

Reports on the first three years of experience under the Act failed to support a number of the dire predictions made by its opponents, such as that the poor, the uninsured, and those without access to palliative care would disproportionately resort to PAS. The data accumulated thus far indicates that the number of patients who actually utilize PAS is quite small, and that they reflected those in the population who died of similar underlying illnesses.[46]

Whether further experience under the Oregon Act will be obtained, and whether other states may choose to follow Oregon's lead, may well depend on the success of the opponents of PAS in the nation's capitol. In the last three years separate (but identical) bills have been introduced in the House and Senate of the Congress of the United States specifically intended to overturn the Oregon Act and prevent similar laws from being implemented in other states. Initially entitled the Lethal Drug Abuse Prevention Act of 1998, and subsequently given the more beneficent title Pain Relief Promotion Act of 1999 (PRPA), the purpose of the legislation was to amend the federal Controlled Substances Act[47] so as to prohibit physicians from "dispensing or distributing or administering a controlled substance for the purpose of causing death or assisting another person in causing death."[48] This prohibition is intended to be unaffected by any state law (such as Oregon's) regulating PAS. PRPA was introduced after Attorney General Janet Reno made the determination that physicians who provided patients with PAS in compliance with the Oregon law would not be in violation of the Controlled Substances Act. In other words, writing a prescription for a lethal controlled substance at the request of a dying patient is a legitimate medical purpose.

PRPA was not enacted into law in either 1999 or 2000. Whether it will be reintroduced in 2001 or thereafter is unknown. The new Republican Attorney General may well reverse the prior administrative determination, thereby threatening any physician who provides PAS to a patient, even in compliance with state law, with loss of prescribing authority and possible criminal prosecution. Even those groups that have opposed PAS have generally opposed PRPA. Their primary reason for doing so is the firm conviction that it will have a profoundly negative impact on the provision of palliative care, inasmuch as the determination of whether a physician's intent in prescribing any controlled substance was to relieve the patient's pain or to cause (or assist in causing) the patient's death would be made by law enforcement agencies. PRPA's supportive words for relieving the pain of dying patients add nothing of consequence to existing law, inasmuch as the Controlled Substances Act does not preclude the provision of opioids to dying patients as medically appropriate to relieve their suffering. The meager amount of funds PRPA allocates to improving end-of-life care are *de minimis* by any standards, and thus serve only to reinforce the true purpose of the bill.

The debate over PAS bears many similarities to the debate over abortion. Both debates have been long on slogans and soundbites and short on sustained public dialogue and close critical analysis, at least outside of journals and texts of law and philosophy. It is the "culture of life" versus the "culture of death." In each instance, a majority view of the population that physicians should be able to provide a particular medical intervention to a patient under carefully controlled circumstances has been denounced by the opposition as murder. The inter-relationship between medicine, law and ethics is vividly demonstrated in each instance. Finally, dire predictions about the slippery slope have been proven not merely exaggerated but patently false. Yet none of this has moderated the opposition, in large measure because of the religious, or at least quasi-religious deontological orthodoxy of their position.[49] If human life begins at conception, then abortion at any stage of pregnancy for any reason is murder. If all human life is sacred regardless of its quality, then hastening death is murder or suicide, and cannot be justified even as a release from great suffering and an imminent, inevitable death from a terminal condition.

11. DISTINGUISHING AGGRESSIVE PALLIATIVE MEASURES FROM PHYSICIAN-ASSISTED SUICIDE AND VOLUNTARY ACTIVE EUTHANASIA

The concurring opinions in *Glucksberg and Quill* sought to reassure health care institutions and professionals that while statutes such as those in Washington and New York were not unconstitutional, criminalizing physician-assisted suicide was not inconsistent with a public policy supporting the provision of aggressive pain relief to dying patients. In this regard, the Court was mindful of the voluminous literature documenting the pervasive and persistent problems in the care of dying patients in the United States.[50] In its effort to distinguish palliative measures that may involve a risk of hastening the patient's death from a lethal prescription or injection, the Court, with precious little argumentation or analysis, invoked the Doctrine of Double Effect (DDE).

DDE is an obscure, ambiguous, and controversial artifact of medieval Catholic theology. It has no previously recognized status in the law, and it has been either ignored or severely criticized in the philosophical literature.[51] Indeed, DDE is in conflict with the traditional legal position on intent, as we shall see. The purpose of DDE was, ostensibly, to provide a means of morally assessing an action that has both a good and a bad effect. Despite the bad effect, according to DDE the action may nevertheless be morally acceptable if the following criteria are met:

- the act itself is either good or morally neutral
- only the good effect is intended
- the good effect is not caused by the bad effect
- the good effect is morally proportionate to the bad effect

While each of the four criteria are controversial, it is the second which is most problematic from a legal perspective, inasmuch as DDE accepts the proposition that the bad effect may be foreseeable but nonetheless unintended. The law, to the contrary, presumes that a person intends the natural and probable consequences of his or her actions The reason for this presumption is that intent is a subjective state of mind that cannot be definitively determined by a third party. Since intent is an essential element of many criminal offenses, the prosecution's burden of proof could never be met unless such a presumption were in place.

As a number of commentators have pointed out, intent in the care of dying patients is an ambiguous phenomenon.[52] When a patient is near death and experiencing considerable, perhaps intractable suffering, the physician (and the patient) may perceive death as preferable to the prolongation of suffering. Moreover, aggressive palliative measures, such as the prescribing of high doses of opioid analgesics, may be virtually indistinguishable from measures that might be undertaken with the intent of causing or hastening the patient's death.[53]

The issue is further complicated by the Court's acceptance of the practice of terminal sedation as morally and legally appropriate for dying patients with intractable suffering. Terminal sedation involves rendering the patient permanently unconscious with the understanding that life-sustaining measures will then be withheld or withdrawn and the patient allowed to die. Opponents of PAS offer terminal sedation as a morally acceptable alterative, while many advocates of PAS maintain that there is no legitimate moral distinction between the two. The latter suggest that it is nothing more than self-deception to maintain that the physician who provides terminal sedation foresees, but does not intend, the patient's death.[54]

The question of whether one can make a plausible and principled moral and legal distinction between intending and foreseeing is of more than merely academic interest because of recently proposed federal legislation discussed at length in Section 10 of this Chapter, as well as the threat of criminal prosecution (however remote). With regard to the threat of criminal proceedings, research indicates that physicians are rarely prosecuted and even more rarely convicted.[55] But perception is reality insofar as motivating physician conduct. Such studies and Justice O'Connor's reassurances in *Glucksberg* to the contrary notwithstanding, physicians acknowledge erring on the side of pain rather than pain relief out of fear of regulatory scrutiny, disciplinary action, and criminal prosecution.

12. THE ROLE OF BIOETHICS AND THE LAW IN ADDRESSING PROBLEMS IN END-OF-LIFE CARE

For the most part, the cases we have considered in the areas of informed consent, advance directives, and refusal of life-sustaining procedures directly by patients or through duly designated surrogates have drawn only narrowly and superficially from the relevant philosophical and bioethical literature. There are a few quite notable exceptions, such as *Bouvia* and the dissents of Justices Brennan and Stevens in *Cruzan*. Several features distinguish these exceptions from most of the cases in general. First, they unhesitatingly confront the nature and extent of the patient's condition and its implications for the life of that patient. Second, they do not concern themselves, and properly so, with future hypothetical patients who are not before the court, or with health care professionals or the state, as though the interests of the latter could ever seriously be considered to trump those of a patient when confronting grave or terminal illness. What rings through these opinions, while largely being silenced or ignored in the others, is the voice of the patient.[56] Third, they appear to comprehend, as the others clearly do not, the concept of a fate worse than death. In the face of significant, intractable suffering, or terrible indignity, they acknowledge that some patients may choose death. But the choice of death in such instances, Justice Scalia's views to the contrary notwithstanding, has nothing whatsoever to do with suicidal ideation, and everything to do with respect for individual autonomy (actual or prospective). The most recent and dramatic failure of American appellate courts to confront the ultimately moral and philosophical issues inherent in human mortality is to be found in the Supreme Court PAS decisions. As one commentator has observed, what reigned on the Court in these cases was a profound and disturbing "disorder."[57] This disorder is attributable, in large measure, to a failure, indeed a refusal on the part of the Court to confront the constitutional implications of human suffering and death. In Chapter 9, we will consider at greater length the place of and the potential role for philosophical and bioethical analysis in judicial decision making.

In the last ten years unprecedented attention has been given to the marked deficiencies in pain management generally and end-of-life care in particular. Many states have formed commissions to study the problem and propose remedial measures, and national organizations such as the Institute of Medicine, the American Board of Internal Medicine, and the American Medical Association have identified core competencies that all physicians should possess. Clinical practice guidelines have been promulgated by federal agencies, national professional organizations, and state medical licensing boards, all intended to improve the quality of pain management and palliative care provided to patients. The question remains, however, whether all of this activity constitutes mere rhetoric or major reform. While we have begun to see a few strongly-worded ethical admonitions to physicians reinforcing the ancient responsibility of the profession to relieve suffering, it nevertheless remains an inescapable fact that physicians are generally at risk only if they over-prescribe pain medications. The exceedingly few instances in which physicians have run afoul of the law for under-prescribing pain medications or in other ways being unresponsive to the pain and suffering of their patients strongly suggests the answer to the rhetoric versus reform question.[58] The same is true of advance care planning and the utilization of and respect for written advance directives. The medical literature is replete with attempts at moral suasion, i.e., ethical admonitions to physicians to talk with their patients about the care they wish to receive, and not receive, as they approach death or face grave and potentially terminal illness. Yet neither ethical nor legal sanctions are currently in place when health professionals fail to ascertain or respect patient wishes. While the PSDA has one sledgehammer sanction for institutions, banishment from the Medicare and Medicaid programs, it has never been imposed, nor has the DHHS even sought to apprise itself of the extent to which institutions are in compliance.

The debate continues to rage over the most appropriate and effective means of modifying the behavior of health care institutions and professionals so as to elevate the prevailing standard of care for pain management and palliative medicine. Neither education nor moral suasion has brought about demonstrable improvement. For example, the assertion in one of the early studies documenting the routine undertreatment of pain that no patient should have to wish for death in order to escape severe pain is echoed in an article sixteen years later insisting that to allow a patient to experience unbearable pain and suffering is unethical medical practice.[59] Yet six years after the second article SUPPORT demonstrated that at least half of all ICU patients in the study died in moderate to severe pain. One senses from reading the clinical literature on the subject that the ethical dimensions of under-treated pain and unaddressed suffering are rarely acknowledged.[60] The instances just cited are the exceptions that confirm the general rule. In the concluding chapter, we shall revisit the issue of the role of law in bringing about major change in social and professional practice.

ENDNOTES AND REFERENCES

1. Robert Veatch, *Death, Dying, and the Biological Revolution* (New Haven: Yale University Press, 1976), p. 25.

2. Report of the Ad Hoc Committee of the Harvard Medical School to Examine the Definition of Death: A Definition of Irreversible Coma, *Journal of the American Medical Association* (1968), 205: 337-340.
3. Henry K. Beecher, "The New Definition of Death – Some Opposing Views," in R. Veatch, *Death, Dying and the Biological Revolution* (New Haven: Yale University Press, 1976), at p. 39.
4. Uniform Determination of Death Act, § 1, 12A U.L.A. 593 (1996).
5. In re Bowman, 617 P.2d 731 (Wash. 1980) (*en banc*).
6. President's Commission for the Study of Ethical Problems in Medicine and Biomedical and Behavioral Research: Defining Death – Medical, Legal, and Ethical Issues in the Determination of Death, (Washington, D.C.: U.S. Gov't. Printing Office, 1981), at p. 42.
7. President's Commission, note 6, p. 35.
8. Roland Pucetti, "Does Anyone Survive Neocortical Death?" in Richard Zaner, ed., *Death: Beyond the Whole-Brain Criteria* (Dordrecht: Kluwer Academic Publishers, 1988), p. 84.
9. Pucetti, note 8.
10. Paul Edwards, "My Death," in Paul Edwards, *The Encyclopedia of Philosophy* (New York: The Macmillan Co, 1967), Vol. 5, p. 416.
11. H. W. Johnstone, Jr., "Sleep and Death," *The Monist* (1976), 59: 218-233, p. 221.
12. John Martin Fischer, "Death," in Lawrence C. Becker, Charlotte B. Becker, *The Encyclopedia of Ethics* (New York: Garland Publishing Co., Inc, 1992), Vol. 1, pp. 239-241.
13. N.J. Stat. Ann. § 26: 6A-5; N.Y. Codes, Rules and Regs., Title 10, § 400-16.
14. President's Commission, note 6, pp. 55-56.
15. Application of the President and Directors of Georgetown College, Inc., 331 F.2d 1000 (D.C. Cir. Ct. App. 1964).
16. For a more detailed discussion of the justiciability issue in refusal of treatment cases, *see* Ben A. Rich, "The Assault on Privacy in Healthcare Decisionmaking," *Denver University Law Review* (1991); 68: 1-55 at 17-21.
17. *See, e.g.,* Public Health Trust of Dade County v. Wons, 541 So.2d 96 (Fla. 1989).
18. Natanson v. Kline, 350 P.2d 1093 (1960).
19. Schloendorff v. The Society of New York Hospital, 105 N.E. 92 (1914).
20. Bouvia v. County of Riverside, California Superior Court, December 16, 1983.
21. Superintendent of Belchertown State School v. Saikewicz, 370 N.E.2d 417 (1977).
22. Ben A. Rich, "The Assault On Privacy In Healthcare Decisionmaking," *Denver University Law Review* (1991), 68: 1-55.
23. Statements attributed to Bouvia's caregivers at the time of this legal dispute raise serious questions as to whether their primary concern was beneficence (doing good for the patient) or self-interest (avoiding legal liability or negative publicity). For example, one of her attending physicians asserted: "When it comes to criminal charges, wrongful death is more of a crime than battery, so there you have it." Of course, if the court had ruled in Bouvia's favor and she died as a result of her treatment refusal, the physician could have had no potential liability arising from his compliance with a court order.
24. Bouvia v. Glenchur, 255 Cal. Rptr. 296 (L.A. Superior Court, 1986).
25. The court, in making this point, moves well beyond a dispassionate articulation of law and policy: "We do not believe it is the policy of this State that all and every life must be preserved against the will of the sufferer. It is incongruous, if not monstrous, for medical practitioners to assert their right to preserve a life that someone else must live, or, more accurately, endure, for '15 to 20 years.' We cannot conceive it to be the policy of this state to inflict such an ordeal on anyone." Bouvia v. Superior Court, 179 Cal. App. 3d 1127 (Second District, 1986).
26. This point closely tracks the language of Justice Burger in the *Georgetown College* case quoted in the text accompanying note 16.

27. For a detailed discussion and analysis of the case of Dax Cowart *see* Lonnie Kliever, ed., *Dax's Case – Essays in Medical Ethics and Human Meaning* (Dallas: SMU Press, 1989).
28. Cruzan v. Harmon, 760 S.W.2d 408 (Mo. Banc 1988).
29. 760 S.W.2d 408, 412 (1988).
30. Cruzan v. Director, Missouri Dept. Health, 110 S. Ct. 2841, 2863-4 (1990), Brennan, J., dissenting.
31. 110 S. Ct. 2841, 2878 (1990).
32. 110 S. Ct. 2841, 2879 (1990), Stevens, J., dissenting.
33. Such a mechanism might be a statutory requirement that any health care institution or professional must notify a designated state office whenever terminating life support is under consideration. One can argue that in the absence of such a process, there is a violation of equal protection of the citizens of Missouri, inasmuch as some of them must contend with the assertion of the state's countervailing interests while many others do not. The reason why states such as Missouri have failed to institute such mechanisms seems quite clear. First, they would be extremely expensive and likely to clog the courts. Second, the citizens of Missouri might be outraged by such a "Big Brother" approach to what are, in theory and usually in practice, private medical decisions.
34. Robert J. Lifton, *The Nazi Doctors* (N.Y.: Basic Books, 1986).
35. 79 F.3d 790 (9th Cir. 1996).
36. Quill v. Vacco, 80 F.3d 716 (2nd Cir. 1996).
37. Vacco v. Quill, 138 L. Ed. 2d 834, 841 (1997).
38. A facial challenge to a statute must persuasively demonstrate that there are no conceivable circumstances in which the statute could be applied without violating some provision of the Constitution. An as applied challenge need only persuasively demonstrate that as applied to the individuals in question, the statute violates the Constitution.
39. The challengers in *Glucksberg* and *Quill* were comprised of two categories. First, competent, terminally ill patients who wished their physician's assistance in hastening their death because the state-of-the-art palliative measures provided to them had been insufficient to reduce their levels of pain and suffering to an acceptable level. Second, the physicians of these patients who, but for the threat of criminal prosecution posed by the statute, would have provided the assistance in dying desired by their patients. These challengers had clearly and consistently maintained that the constitutional infirmity of the statutes in question was only *as applied* to them and other similarly situated competent, terminally ill patients.
40. Washington v. Glucksberg, 138 L. Ed. 2d 772, 798 (1997), O'Connor, J., concurring.
41. For just two examples from this literature, that in turn cite many others, *see* Kathleen M. Foley, "The 'Decriminalization' of Cancer Pain," *Adv. Pain Res. & Therapy* (1989), 11: 5-17; Phoebe S. Haugen, "Pain Relief for the Dying: The Unwelcome Intrusion of the Criminal Law," *William Mitchell L. Rev.* (1997), 23: 325-375.
42. Washington v. Glucksberg, 138 L.Ed.2d 772, 833 (1997), Breyer, J., concurring.
43. Robert A. Burt, "The Supreme Court Speaks: Not Assisted Suicide but a Constitutional Right to Palliative Care," *New England Journal of Medicine* (1997), 337: 1234-1236.
44. The first Constitutional challenge to such a law may be launched against a bill that has recently passed both houses of Congress with the significantly misleading title of "The Pain Relief Promotion Act of 1999" (PRPA). While disclaiming any intent to discourage the provision of appropriate palliative care to dying patients, the act make it a federal criminal offense to prescribe or administer controlled substances (e.g., Schedule II narcotics such as morphine) to a patient with the intent of causing or assisting in the death of a patient. Whether the intent of the physician was to manage pain or cause death would be left to the determination of law enforcement officials, with such input from medical experts as they choose to solicit. The clear intent of the Act is to nullify the Oregon Death With Dignity Act (legalizing and regulating physician-assisted suicide). Both Compassion In Dying, which

initiated *Glucksberg* and *Quill*, and the Oregon Attorney General's Office, have made it clear that if PRPA becomes law, they will initiate Constitutional challenges to it.

45. Oregon Rev. Stat. §§ 127.800-.897 (1995). Indeed, the Oregon legislature was so heavily influenced by well-organized and generously funded right-to-life organizations that it returned the issue to the electorate for a second vote. Whereas the referendum's margin of victory had been a mere 51% to 49% in 1994, it increased to 60% to 40% in 1997. The shift is largely attributed to voter outrage that the legislature had forced them to address the issue a second time.
46. Arthur E. Chin, et al., "Legalized Physician-Assisted Suicide in Oregon – The First Year's Experience," *New England Journal of Medicine* (1999), 340: 577-583.; Amy D. Sullivan, Katrina Hedberg, David W. Fleming, "Legalized Physician-Assisted Suicide in Oregon – The Second Year," *New England Journal of Medicine* (2000), 342: 598-604; Oregon Health Division - Center for Health Statistics, "Oregon's Death With Dignity Act: Three Years of Legalized Physician-Assisted Suicide," http://www.ohd.hr.state.or.us/chs/pas/arresult.htm.
47. 21 U.S.C. § 823.
48. H.R. 2260 (S. 1272), Pain Relief Promotion Act.
49. Not all of physician-assisted suicide's harshest critics are from the ranks of organized religion. *See* Willard Gaylin, Leon Kass, Edmund Pellegrino, and Mark Siegler, "Commentary: Doctors Must Not Kill," *Journal of the American Medical Association* (1988), 259: 2139-2140; Carlos Gomez, *Regulating Death: Euthanasia and the Case of the Netherlands* (N.Y.: Free Press, 1991); Herbert Hendin, *Seduced by Death: Doctors, Patients, and the Dutch Cure* (N.Y.: W.W. Norton, 1997); Yale Kamisar, "Are Laws Against Assisted Suicide Unconstitutional?" *Hastings Center Report* (1993), 23: 32-39.
50. For a detailed discussion of medicine's legacy of undertreated pain *see* Ben A. Rich, "A Prescription for the Pain: The Emerging Standard of Care for Pain Management," *William Mitchell Law Review* (2000), 26: 1-91.
51. Phillipa Foot, "Abortion and the Doctrine of Double Effect," in *Virtues and Vices and Other Essays in Moral Philosophy* (Berkeley: University of California Press, 1978), pp. 19-32.
52. Timothy Quill, "The Ambiguity of Clinical Intentions," *New England Journal of Medicine* (1993), 329: 1039-40.
53. Norman L. Cantor and George C. Thomas, "Pain Relief: Acceleration of Death and the Criminal Law," *Kennedy Institute of Ethics Journal* (1996), 6: 107-127.
54. David Orentlicher, "The Supreme Court and Physician-Assisted Suicide: Rejecting Assisted Suicide But Embracing Euthanasia," *New England Journal of Medicine* (1997), 337; 1236-1239.
55. Ann Alpers, "Criminal Act or Palliative Care? Prosecutions Involving the Care of the Dying," *Journal of Law, Medicine & Ethics* (1998), 26: 308-331.
56. For an extensive feminist analysis of this dimension of the Supreme Court decisions on abortion and physician-assisted suicide *see* Susan M. Behuniak, *A Caring Jurisprudence* (New York: Rowman & Littlefield Publishers, Inc., 1999).
57. Robert A. Burt, "Disorder in the Court: Physician-Assisted Suicide and the Constitution," *Minnesota Law Review* (1998), 82: 965-981.
58. Despite the massive amount of data indicating that instances of under-prescribing opioid analgesics for pain greatly exceed instances of over-prescribing, thus far only one state medical board (Oregon in 1999) has taken disciplinary action against a physician for subjecting patients to unnecessary suffering. Furthermore, only a single case has been noted in which a health care provider was adjudged to have negligently failed to properly manage a patient's pain, and that case did not involve a physician but rather a nursing home and one of its nurse employees. Estate of Henry James v. Hillhaven Corp., No. 89 CVS 64 (N.C. Super. Ct., Jan. 15, 1991).

59. R. M. Marks and E. J. Sachar, "Undertreatment of Medical Inpatients With Narcotic Analgesics," *Annals of Internal Medicine* (1973), 78: 173-181; Sidney H. Wanzer, et al., "The Physician's Responsibility Towards Hopelessly Ill Patients – A Second Look," *The New England Journal of Medicine* (1989), 320: 844-849.
60. Ben A. Rich, "A Legacy of Silence: Bioethics and the Culture of Pain," *J. Medical Humanities* (1997), 18: 233-259; Ben A. Rich, "An Ethical Analysis of the Barriers to Effective Pain Management," *Cambridge Quarterly of Healthcare Ethics* (2000), 9: 54-70.

Chapter 9

THE NEW SYNERGY – BIOETHICS IN COURT

1. INTRODUCTION

In Chapter 3 we considered the traditional debates over the relationship between law and morality. In this chapter we revisit that debate in the contemporary context of bioethics. The nature of the debate in this context is somewhat different, as well as noticeably more complex and nuanced. There are several elements that we need to consider. The first, which has the most direct lineage with the historical debate, is whether ethical theory in general, but particularly as articulated by academic moral philosophers of all stripes and shades, including bioethicists, has anything of merit to contribute to the analysis and resolution of legal cases with ethical dimensions or implications. This element has been quite recently and contentiously raised by the prolific federal appellate court judge Richard Posner. Judge Posner has a distinctly skeptical view of the contributions that academic moral philosophy can make to the adjudication of cases, but his critics (and the objects of his scorn) are some of the most prominent American moral and political philosophers, including Ronald Dworkin, Martha Nussbaum, John Rawls, and Judith Jarvis Thomson.

A second element is actually a concern, most often expressed by health professionals and non-lawyer bioethicists, that in the United States law has essentially dictated, or at least profoundly influenced, the ways in which we consider and resolve bioethical issues. This apparent subsuming of bioethics into the body of law in a particular branch of medical jurisprudence is considered by some to be both a problem and a threat to bioethics more broadly considered. However, whether there is any genuine substantive basis for this concern, or whether it merely reflects a more general distaste for lawyers and legislators, remains to be seen.

The third element is the increasingly activist role that bioethicists (attorneys and non-attorneys alike) are playing in the resolution of legal disputes. There are actually two roles at issue – serving as expert witnesses in trial court proceedings, and participating in the preparation and submission of *amicus curiae* (friend of the court) briefs to appellate courts, including most notably the United States Supreme Court. There has recently developed a debate in the bioethics literature as to whether either of these roles, but most significantly the former, are appropriate.

2. POSNER ON THE PROBLEMATICS OF MORAL AND LEGAL THEORY

Judge Richard Posner compiled his theses and arguments in opposition to the notion that courts should concern themselves with the work of academic moral theorists in the process of deciding cases with significant moral dimensions in a recent book.[1] However, because Posner unabashedly considers himself to be following in the tracks of that famous jurist of another era, Oliver Wendell Holmes, any consideration of Posner's arguments, however attenuated, must start with the seminal law journal article by Holmes in which the basic proposition was originally formulated.[2] At the heart of Holmes' view was the idea that to be effective law must command obedience, if not respect, by all those in society – the moral, the amoral and the immoral. Thus, while Holmes acknowledged that "[t]he law is the witness and external deposit of our moral life" and that "[i]ts history is the history of the moral development of the race," he goes on to say that "[i]f you want to know the law and nothing else, you must look at it as a bad man, who cares only for the material consequences which such knowledge enables him to predict, not as a good one, who finds his reasons for conduct . . . in the vaguer sanctions of conscience."[3] Much confusion has arisen, according to Holmes, because law and morality share so many features, from a common vocabulary to common goals of influencing human behavior in ways that contribute to the well-being of society.

Holmes also believed, as the quintessential legal pragmatist whose direct descendent is the eminent Judge Posner, that morality is concerned with the "actual internal state of the individual's mind, what he actually intends" whereas the law is only concerned with inducing compliance with its provisions. This is a bit of an overstatement, since there are certainly criminal offenses in which intent, or *mens rea* (guilty mind), is an essential element that the state must prove. However, generally Holmes is correct that it is, for all practical purposes, a thing indifferent to the law whether someone has developed a habit of complying with the law because they believe in their heart that it is the "right" thing to do, akin to the Kantian notion of the categorical imperative, or simply because they have performed their own personal hedonic calculus and decided that the costs of disobeying the law outweigh any potential benefits. For Holmes, a legal duty is "nothing but a prediction that if a man does or omits certain things he will be made to suffer in this or that way by judgment of the court."[4]

In Chapter 10 we will concern ourselves in significant part with another fundamental distinction between law and morality that is not developed in Holmes' article – the contemporary disparity between moral (ethical) and legal sanctions. One of the strongest arguments in support of addressing bioethical issues through the law is that the law has the sanctions often required to compel respect for such things as patient autonomy, whereas mere ethical admonitions, even from state or national medical organizations, can be and often are ignored with impunity. We considered at length a prime example of this in Chapter 8 – the SUPPORT data. It is hard to believe that 50% of all ICU patients would have died in pain and had their wishes with regard to treatment ignored if their physicians had faced genuine legal sanctions for such conduct. The challenge for bioethics, at least insofar as it seeks to diminish the extent to which it has been subsumed under medical jurisprudence, is to show how it is possible to ensure compliance with ethical principles without resort to the courts and or the legislatures.

Posner espouses a position of "pragmatic moral skepticism," the essence of which is a denial that there are "universal moral laws ontologically akin to scientific laws" or that the rational arguments of moralists have the power to persuade people to change "their beliefs and the behavior the beliefs might influence."[5] However, for purposes of our discussion, Posner's most important claim is that judges do not confront moral issues when they decide cases, regardless of the nature of the issues before them, and therefore they do not need to take moral theory into account in the adjudication process. We can thus see how Posner's thesis has significant implications for another topic of this chapter, the bioethicist in court. One of the things that enables Posner to cavalierly contend that in deciding cases like *Roe v. Wade, Planned Parenthood of Southeastern Pennsylvania v. Casey, Cruzan v. Director, MO Dept. of Health,* and *Washington v. Glucksberg* the Supreme Court was not drawing upon or applying moral principles is his extremely narrow and compartmentalized view of the moral domain. While some constitutional law scholars might be willing to concede that those decisions did not directly turn on some precise principle of bioethics, few, I suspect, would agree with Posner that those decisions did not draw upon and reflect political morality.[6] While Justice Blackmun did assert in *Roe* that the Court was not purporting to answer the profound philosophical question of when human life begins, it most certainly was deciding a fundamental question of political morality – that a fetus is not a person in the constitutional sense. *Casey*, at least according to the joint opinion by Justices Kennedy, O'Connor and Souter, was affirming the core holding of *Roe*, and in doing so invoked the language quoted at length in Chapter 7, about the constitutional protections afforded "matters involving the most intimate and personal choices a person may make in a lifetime, choices central to personal identity and autonomy." When the opinion continues by asserting that "[a]t the heart of liberty is the right to define one's own concept of existence, of meaning, of the universe, and of the mystery of human life," Posner's argument that in all of these cases the Court "ducked the philosophical issue" becomes implausible in the extreme. Even in *Cruzan*, where one sympathetic to Posner's position might choose to focus on Chief Justice Rehnquist's Opinion of the Court narrowly holding that Missouri's clear and convincing evidence standard for surrogate decision making did not contravene the Constitution, he also grudgingly conceded that a Fourteenth Amendment liberty interest in refusing life-sustaining medical interventions could be inferred from the Court's prior holdings. The claim that a competent patient has the right to refuse such interventions is an unqualifiedly moral one. Finally, as we have previously noted, in *Glucksberg* and *Quill* five of the nine justices strongly suggested in concurring opinions that dying patients may have a constitutionally protected interest in receiving whatever pain relief they require, even if it posed the risk of a hastened death. Such an interest would necessarily be grounded on a political morality that protects the vulnerable and a bioethics that posits the duty to relieve suffering (as a core value and fundamental principle of medicine).

There is a sense in which Posner's critique of moral philosophy, and his assiduous efforts to insulate jurisprudence from it, are derived from unreasonable expectations. It is as though Posner is claiming that if moral philosophy in general, and academic moral philosophers in particular, were worth anything they would have first developed their own enduring consensus on such contentious issues as the morality of abortion and euthanasia, and then they would have moved out of their respective ivory towers and ivied quadrangles and produced a broad public consensus on these important questions. Having demonstrably failed to do so, Posner argues that judges should ostracize them from the precincts of the courts on the grounds that they have nothing meaningful to contribute. One might just as easily cast the blame right back at his brethren in the courts. Like his hero Holmes, Posner considers certainty to be a cardinal virtue. Since the moral philosophers have not brought certainty to profound moral issues, they are worthless. Assuming for the sake of argument that Posner's contention that the great "bioethical" decisions of the Supreme Court discussed above are actually devoid of moral content, why have these decisions failed to resolve the intractable public debates over these issues? If certainty is the cardinal virtue of the law, and that virtue can be achieved without concern for questions of morality, then why have *Roe* and *Casey* not brought certainty to the abortion issue, nor *Cruzan* and *Quill* certainty to the end-of-life issues? Only if the justices had failed to "duck the moral issues" might Posner's critique strike home. As it is, the law has wrought, or at least failed to remove, as much uncertainty as moral philosophy and its applied branch, bioethics.

Posner's prime adversary in the debate over the place, if any, of moral philosophy in law, and particularly in common law and constitutional jurisprudence, is Ronald Dworkin. As we noted in Chapter 3, Dworkin believes that a "moral reading of the Constitution" is absolutely essential because the Bill of Rights is, in his estimation, a set of broad and abstract principles of political morality. Many sections of the original Constitution are precise and unambiguous in their delineation of the structure of the federal government, such as Article I, Section 3 providing for two senators from every state or Article III, Section 1 vesting the judicial power of the United States in one Supreme Court and such inferior courts as Congress may establish. It is quite plausible to maintain that such provisions need only be referenced and applied, not interpreted. Moreover, they are not imbued with profound moral implications. Some of the most important provisions of the Bill of Rights, however, such as the First Amendment's reference to freedom of speech and of the press, or the Fourteenth Amendment's reference to due process and equal protection of the law, require considerable interpretation. Furthermore, the deprivations of life, liberty, and property from which persons are protected by the Due Process and Equal Protection clauses, are matters of considerable moral consequence. Indeed, those who drafted and adopted the Constitution and its original Bill of Rights were significantly influenced by the philosophical ideas of John Locke and other Enlightenment philosophers who firmly believed in the existence of natural rights of persons that transcended any and all governments. To suggest that these natural rights were devoid of any normative dimensions, and did not require interpretation in subsequent centuries and radically different circumstances, is to fundamentally misapprehend their history and significance.

3. THE INFLUENCE OF LAW ON AMERICAN BIOETHICS

There is no dispute whatsoever about the fact that, at least in the United States during the last thirty years, law has had an immense influence upon the ethics of medicine. There is a significant dispute, however, as to whether, on balance, the nature of that influence has been positive or negative. We ought not to be surprised that a majority of physicians and others in the various health professions are quite derogatory about the role of law, and more particularly its practitioners, in bringing about a state of affairs that they find extremely problematic. In the good old days, before, let us say, the 1950's, the law and lawyers knew their place, and physicians reigned supreme in medical practice and research. While there might be the occasional medical malpractice suit, the fact remained that most physicians could retire from many decades in the practice of medicine without ever having to contend with a lawyer unless he chose to do so, as in the drafting of a will or other aspects of estate planning. As I noted in Chapter 1, however, there is at least one lawyer (law professor) who has written at length about the generally pernicious role that, in his view, law has had on medicine.[7] There is a hint of a Posnerian critique in Dworkin's writing, for he faults law for bringing a nonscientific approach to what he considers to be an essentially scientific enterprise, i.e., the diagnostic and therapeutics of medical practice and, of course, medical research. What this critique overlooks, or at least insufficiently acknowledges, however, is that medicine has been viewed since its inception, and by some of its most distinguished practitioners, as fundamentally a moral enterprise.[8] Indeed, Robert Veatch, the author of an earlier volume in the medical ethics series in which Dworkin's book appears identified as the most salient feature of what he characterized as the new era of "postmodern medicine" the debunking of the longstanding myth that medicine was pure science and hence devoid of any normative dimensions. Quite to the contrary, Veatch asserted, virtually every aspect of medicine is fraught with normative implications.[9] Indeed, the very title of Veatch's book, *The Patient As Partner*, suggests the most positive influence that law has had on medicine – the focus of Chapter 4 of this book – the transformation of the paradigm of patient care from that of paternalism to that of shared decision making. Remarkably, in the process of denigrating the role of law in biomedicine and bioethics, Dworkin does not even mention the legal origins of the doctrine of informed consent or its transformative influence on the nature of the relationship which is at the heart of the medical encounter.[10]

In stark contrast to the views of Roger Dworkin are those of George Annas. The latter does not see the influence of law upon medicine as one of pernicious intermeddling. Rather, he asserts that while it may be somewhat of an exaggeration to claim that "American bioethics is engaged in a struggle for the soul of American medicine . . . It is not too much to say that the values and ideals of American law, as expressed in the Declaration of Independence and the Constitution, have consistently dominated the ethical discourse concerning both substantive and procedural approaches to medical practice."[11] Those values and ideals, of course, are epitomized by human liberty and the right to self-determination. Such values were nowhere within the panoply of ideals sacred to medicine, as we noted in Chapter 1, at least outside of the domain of physician autonomy and self-determination.

In the section that follows we will consider ways in which bioethics may seek to assert more of an influence over law. Then, in Chapter 10, we will consider whether bioethics, as presently constituted, has the capacity to take over from law the predominant position in shaping the normative dimensions of medical practice and research, as well as the moral implications of national health policy or the lack thereof. One thing should be quite clear by this point in our analysis – this book is not intended to be an unqualified panegyric on how law, particularly the courts, have comported themselves when confronting medical issues. We have considered our share of cases in which the only positive statements pertain to dissenting opinions, or in which indefensible rulings by trial judges had to be rectified at the appellate level. But the felt need of parties on both sides of ethical disputes – health care institutions and professionals on the one hand and patients and families on the other – to seek judicial involvement reflects a lack of shared values in health care decision making. As Alex Capron rightly observed: "In a society in which ethical standards were sufficiently complete to address even novel technical problems, widely enough shared to be accepted by all or nearly all persons, and consistent and coherent enough never to lead to uncertain or contradictory results, bioethics might operate with little reference to the law."[12] But as we have seen, even ethical admonitions by the American Medical Association in its *Code of Medical Ethics* have been shown by studies such as SUPPORT (Chapter 8) to be ignored by practitioners in the premier academic medical centers. We seem to be a long way from achieving the shared values in health care that Professor Capron believes may be the antidote to the legalization of medicine and bioethics.

4. THE BIOETHICIST IN COURT

We begin with a discussion of the less controversial of the two means by which bioethicists have sought to influence the outcome of legal cases with bioethical implications – the *amicus curiae* brief. Such "friend of the court" briefs have now become quite routine in high profile cases under review by appellate courts, particularly the federal circuit courts of appeal and the U.S. Supreme Court. George Annas was the organizing force behind the first amicus brief filed in the Supreme Court by a group of bioethicists. The case, noted in Chapter 7, was *Webster v. Reproductive Health Services*.[13] The argument of the brief was that the essential holding of *Roe v. Wade* should not be overruled. Annas observes that the decision to submit such a brief was controversial, in part because of the appearance that bioethics was placing a stamp of approval on the morality of abortion.[14] To the contrary, he contends, offering general support for a woman's *legal* right to choose to terminate a pregnancy prior to viability makes no value judgment about whether it is *ethical* to so in any particular circumstance. When characterized in this way, however, one might wonder what particular expertise the bioethicists bring to the Court. However, such a brief makes clear that respect for procreative liberty does not entail a blanket ethical endorsement of all abortions, nor would a blanket prohibition of abortion (by a state) always produce the most ethical outcome in a particular case.

In the physician-assisted suicide cases, bioethicists and moral philosophers submitted opposing *amicus* briefs. What has come to be characterized as "The Philosophers' Brief" was prepared by six of the most prominent moral and political philosophers in the United States: Ronald Dworkin, Thomas Nagel, Robert Nozick, John Rawls, Thomas Scanlon, and Judith Jarvis Thomson. While openly acknowledging that they differ on many issues of morality and public policy, they are united by a conviction that "respect for fundamental principles of liberty and justice, as well as for the American constitutional tradition, requires that the decisions of the Courts of Appeals be affirmed."[15] In affirming those decisions, they argue, the Supreme Court need not make "moral, ethical, or religious judgments about how people should approach or confront death," but rather that the Court should prevent the states from imposing such judgments on individuals.

The brief makes three specific arguments: (1) the liberty interest asserted here [by the original plaintiffs] is protected by the Due Process Clause; (2) this Court's decisions in *Casey* and *Cruzan* compel recognition of a liberty interest here; and (3) state interests do not justify a categorical prohibition on all assisted suicide. The analysis and discussion in support of each of these three arguments is straightforwardly legal in nature. The nature of the holdings of the Court in related or analogous cases is considered and argued to provide precedents that under the principle of *stare decisis* compel the outcome in these cases for which the brief advocates. Even in argument number 3, which might have been couched as one relying on principles of political philosophy, the authors instead rely primarily on legal precedent. Particular emphasis is placed upon the curious argument of the Solicitor General of the United States in these cases, which conceded that "a competent, terminally ill adult has a constitutionally cognizable liberty interest in avoiding the kind of suffering experienced by the plaintiffs in this case," but that states nevertheless have the right to "override" this interest on the grounds that even the most stringent regulation of the practice of physician-assisted suicide would unreasonably endanger the lives of patients whose desire for such measures is uncertain or is the product of despair or coercion.

Of particular interest, in light of our discussion of the actual Opinions of the Court in these cases in Chapter 8, is the conclusion to "The Philosophers' Brief." The specific circumstances of each of the named patient-plaintiffs in these cases is described and their request for a physician-assisted death noted to be "competent, rational, informed, stable, and uncoerced." Clearly, those philosophers submitting the brief believed that the challenge to the state statutes prohibiting assisted suicide was "as applied" to these particular patients (and others similarly situated), not a facial challenge to any conceivable application of those laws. As we noted in Chapter 8, despite the Ninth Circuit's conclusion that the Washington law was unconstitutional as applied to the named plaintiffs, the Supreme Court treated the challenge as a facial one. In any event, the general thrust of "The Philosophers' Brief" is decidedly legal rather than philosophical in nature. One does not find, as might be expected from the title of the brief, references to and discourses upon major works of moral, political, or even legal philosophy and their implications for the cases before the Court. Neither does one find unique insights into the philosophical dimensions of the cases before the Court that bear upon their adjudication and which only academic philosophers of this stature might be expected to provide. Judge Posner, in his book discussed earlier in this chapter, with apparent glee notes that "The Philosophers' Brief" is not mentioned in either the Opinions of the Court or in any of the concurring opinions in the physician-assisted suicide cases, suggesting (at least to him) that the justices were singularly unimpressed with the arguments it made.

Taking the opposite position before the Supreme Court in these cases was what I shall refer to for purposes of this discussion as "The Bioethicists' Brief." The fifty bioethics professors who signed on to this brief were an interdisciplinary assortment of individuals trained in medicine, law, philosophy, and various other disciplines in the humanities and social sciences. Their stated interest in the litigation was a desire to maintain both a conceptual and practical distinction between a woman's right to an abortion and the right of all competent patients to refuse life-sustaining treatment on the one hand and a purported right to physician-assisted suicide on the other hand. In the view of these *amici*, the Ninth and Second Circuit Courts of Appeal had incorrectly maintained that there is no reasonable distinction between the right to refuse life-sustaining treatment and the right to physician-assisted suicide. In the brief these bioethicists assert that there are significant distinctions between the two that would justify the Court in reaffirming the constitutional liberty interest in refusing life-sustaining treatment while at the same time declining to recognize a right, however limited, to physician-assisted suicide.

Like the Philosophers' Brief, the Bioethicists' Brief relies on legal argument and interpretation rather than a more philosophical approach. A considerable portion of it is devoted to elucidating ways in which its authors believe that the Ninth and Second Circuit Courts of Appeal misinterpreted important cases in medical jurisprudence. In this regard, a number of the arguments in the brief are not only at odds with the views expressed by the Ninth and Second Circuits, but also some of those in Chapters 4 and 8 of this book. For example, the brief asserts that "The Ninth Circuit Court of Appeals mistakenly believes that the right of individuals to refuse treatment is a modern invention that was only recently bestowed upon individuals by the courts." If, as the brief goes on to contend, "[t]he principles of bodily integrity and self-determination are so basic that they seem to need no justification," then how was it that Elizabeth Bouvia came to be forcefed during two separate hospitalizations in California, ultimately having to resort to an appellate court in order to have this "basic" principle recognized and vindicated? Interestingly, the brief states that a rule requiring a competent adult patient to comply with physician orders "would convert medical advice into medical tyranny." That, of course, is precisely how the appellate court in *Bouvia* characterized the conduct of the hospital and physicians, which had been perpetrated with the advice and consent of two trial court judges. As for the bioethicists' contention that "[o]ne searches in vain to find a case in which free living competent adults are restrained against their will and forced to submit to treatment by physicians," one wonders how diligently they searched if the *Bouvia* case is glaringly omitted from the brief.[16]

Perhaps the most notorious instance in which a supposedly free and competent adult was subjected to extended and excruciating treatment against his will is the case of Dax Cowart, to which we briefly alluded in Section 5 of Chapter 8. The bioethics *amici* may dispute that this is a legitimate case, inasmuch as Cowart never succeeded in securing an attorney who would file a legal action on his behalf to challenge the treatment that he was refusing. In other words, Dax never received his day in court, despite diligent and persistent effforts. However, it is quite relevant because their brief seeks to portray a socio-political climate in the United States predating and not reliant in any significant regard on the "right to die" litigation phenomenon of the 1970's and 1980's, in which the medical profession would not have had the temerity to challenge "principles of bodily integrity and self-determination . . . so basic that they seem to need no justification." Later in the brief the *amici* assert: "Any competent adult can refuse treatment for any reason . . . Nor does a patient need the permission or cooperation of a physician to refuse treatment. If a person refused treatment, the physician may not lawfully treat the person. If the patient withdraws consent to treatment, the physician must cease to treat the person." One need only apply this language to "Dax's Case" in order to measure its illusory nature. Many physicians who specialize in the treatment of burn patients will tell you that they presume that their patients are not competent, and operate on a "best interests" approach. Dax Cowart and other physically frail or disabled patients in hospitals (such as Elizabeth Bouvia) do not have the physical ability to get up and leave the hospital when their rejections of a physician's therapeutic belligerence are ignored. The bitter irony of the amici's argument is that only when such patients have prompt and continuous access to the courts (in actuality, not theoretically) do they have any realistic hope of vindicating this right to refuse treatment that the amici seek to reassure us is so much a part of the American social and political culture that no reasonable, law-abiding person (or physician) would even think of abridging it.

According to the "Bioethicists' Brief," one of the fundamental infirmities of the purported right to physician-assisted suicide that the Ninth and Second Circuits recognized in the case of competent, terminally ill patients is the very fact of terminal illness. The *amici* argue that "[t]here is no accepted definition of terminal illness, and it is very difficult, if not impossible, for physicians to accurately predict any particular individual's life expectancy." If any purported right of a patient may be rejected, in whole or in part, because it relies on a terminal prognosis, then much more than physician-assisted suicide is implicated. Every living will statute has a terminal condition requirement. DNR orders generally follow from a physician's determination that the patient is near death, or that their condition is terminal, irreversible, and sufficiently advanced that resuscitation would be futile or is for other reasons contraindicated. Aggressive palliative measures and referral to hospice are almost invariably contingent upon a determination that the patient has only weeks or months to live. If terminal condition is "an undefined and undefinable term," then why did some of these *amici* join in one earlier published admonition that individuals should execute advance directives,[17] or another deploring the care provided to dying patients?[18] If we really cannot expect physicians to recognize a dying patient when they see one, then we really cannot insist that they behave in a particular fashion when they care for such patients. There is something quite disconcerting about the fact that a terminal condition requirement troubled so many bioethicists little or not at all when living wills and appropriate end-of-life care were the issue, but it becomes hopelessly problematic the minute the focus shifts to physician-assisted suicide.

The remaining two sections of the "Bioethicists' Brief" challenge the Ninth and Second Circuits' conclusions that there are significant similarities between withholding or withdrawing life-sustaining treatment and physician-assisted suicide, as well as the Ninth Circuit's view that a woman's right to terminate a pregnancy (with the competent assistance of a physician) implicates and informs a dying patient's right to the assistance of a physician in hastening death. Its authors seek to persuade the Court that one can be, as presumably they all are, staunch defenders of a constitutional right to refuse treatment and to obtain an abortion, without having to accept the Ninth Circuit's proposition that a right to determine the time and manner of one's own death follows naturally, logically, and inevitably from them. The Bioethicists' Brief argues, with varying degrees of persuasiveness, that abortion is a recognized medical procedure but that physician-assisted suicide is not; that regulating physician-assisted suicide would require state intrusion into the privacy of the physician-patient relationship far beyond the regulation of abortion; and that once recognized, a right to physician-assisted suicide could not be limited to patients with a terminal condition.

What should be clear from the discussion of the Philosophers' Brief and the Bioethicists' Brief submitted in the physician-assisted suicide cases is that for the most part such *amicus* briefs are long on traditional legal argumentation and short on more traditional philosophical discourse. This is much more of an observation than a critique. The more such briefs conform to the usual and customary form and content, the less they are subject to the criticism that their proponents are in any sense attempting to instruct the appellate courts on what would constitute a morally acceptable decision. Moreover, such briefs provide confirmation by their authors that legal and bioethical issues are inextricably intertwined in these cases. In other words, no such *amicus* brief has sought to make an argument of the type Roger Dworkin makes in *Limits* that the courts should simply decline to become involved in such cases because to do so would push the courts beyond their limits of competence.

We turn now to the activity in which bioethicists have been most subject to criticism, both from inside and outside their profession – expert testimony in trial court proceedings. Our objective is to understand the nature of the critique and assess its merits.

5. THE BIOETHICIST AS EXPERT WITNESS

Anecdotal evidence suggests that with increasing frequency bioethicists are beings sought by attorneys to testify as expert witnesses in lawsuits raising ethical issues. While it is certainly the case that the parties of record in cases on appeal, or their attorneys, will solicit the filing of *amicus* briefs by organizations or less formally organized groups who support their position on the issues, often those who submit such briefs do so on their own initiative. Interestingly, despite the fact that groups of academic philosophers, physicians or attorneys such as those submitting the Philosophers' and Bioethicists' briefs demonstrate a strong sense of advocacy for their positions, they have not generally been accused of abandoning the disinterestedness that is deemed essential to a true academic. Yet, one of the most frequently raised arguments against bioethicists testifying as expert witnesses at trial is that in doing so they must choose sides in the controversy and hence become advocates with an inclination to slant their testimony in favor of the party who calls them and (presumably) provides their compensation.

One of the perennial criticisms of attorneys as professionals, lodged by those who either do not understand or do not accept the adversary system of adjudication, is that attorneys are merely "hired guns," amoral persons devoid of principles, who will represent anyone who has the ability to pay the price of their representation. By insinuating themselves into such an environment, so the argument goes, the bioethicist must forever abandon any claim to fairness, objectivity, and the requisite academic aloofness from how a particular legal dispute should be resolved. To extend the troublesome metaphor, the bioethicist as expert simply becomes another bullet in the chamber of the hired gun, introduced for the sole purpose and with the specific intent of maiming or fatally wounding the opponent's case. To the extent that this image is at all accurate, it offers a seemingly compelling reason for bioethicists to "just say no" when they hear the siren's song of the attorney in search of bioethical expertise.

Interestingly, a similar admonition has not been made to other academics. Members of the professoriate in medicine, engineering, the hard sciences, and even the social sciences routinely testify as experts in litigation without, it would appear, imperiling their credentials as academics. While there have been physicians who have, quite literally, ceased the active practice of medicine in favor of becoming full-time expert witnesses for hire to plaintiffs' attorneys in medical malpractice cases, they have rarely, if ever, come from the ranks of academic physicians. We must wonder, then, what is peculiar about bioethics that renders its practitioners so exquisitely vulnerable to the "corrupting" influence of trial court litigation.

One suspects that some opponents of the idea that it is acceptable for bioethicists to take on the role of expert witness labor under the mistaken impression that their testimony is for sale, and will be configured or reconfigured to meet the needs of the highest bidder. This perception calls to mind the manner in which trial counsel often cross-examines the opponent's expert witness by asking him or her: "How much are you being paid for your testimony here today?" The inference the jury is expected to draw is that the opinions offered by the expert are tainted, and so too is his or her credibility, by the fact of their purchase. This ploy rarely succeeds, however, because the expert is admonished by the attorney by whom they have been engaged that the appropriate and accurate response is "I am not being paid for my testimony; I am being compensated for my time." Thus, the countervailing inference is that the substance of the expert's opinions has not been fashioned to meet the needs or expectations of the proponent of the witness.

Even some bioethicists who have testified as experts and continue to believe that such a role can be justified see it as necessarily adversarial. One of them, John Fletcher, states: "The core expectation in the expert witness role is that the bioethicist will be an effective advocate of one and only one right moral judgment . . . the courtroom is the place for passionate advocacy of a particular position or moral judgment."[19] So serious are the conflicts between the role of the bioethicist as expert witness and the two other most common roles played by bioethicists – consultant and scholar/teacher – that Fletcher believes certain demanding standards must be met before a bioethicist agrees to take on the role of expert. Among them are that the legal case be "pathbreaking . . . where the stakes for bioethical issues are very high"[20] and that expert witness fees be declined "to avoid harm to . . . reputation."[21] While Fletcher denies passing any moral judgment on bioethicists who follow a less rigid standard, one much more in line with experts in other areas of academia, it is clear that he believes that adherence to the highest of standards is appropriate because of a special, vulnerable status bioethicists currently have in society.

There are, however, plausible counter-arguments to Fletcher's proposed standards. First, one cannot always be blessed with the gift of prescience necessary to determine in advance which legal cases will ultimately prove to be pathbreaking. When the Cruzan family first sought relief in the courts from their daughter's artificial nutrition and hydration, they surely could not have imagined that their case would one day come before the Supreme Court and their name become a household word. There is another concern, which touches both this purported standard and the "no fees" admonition as well. When a bioethicist operates under the impression that the cosmic significance of the cause justifies testimony as an expert witness *pro bono publico* (for the public good), arguably one has ceased to be a witness in a case and instead become a zealous advocate for a cause. Yet in our system of government, the courts exist to decide cases, and the proper forum to pursue a cause lies elsewhere. Thus, rather than serving as some litmus paper test for objectivity, Fletcher's standards start to look more like the indicia that the bioethicist has abandoned objectivity in order to pursue a higher goal.

Some critics of bioethicists as experts maintain that courts are charged with deciding legal disputes, not moral ones. Thus, all bioethics testimony should be subject to exclusion as irrelevant and immaterial. Hopefully, the discussion and analysis in the prior chapters indicate why this objection is untenable. We need to consider only one of the many issues previously considered – the state interests deemed countervailing to the patient's right to refuse treatment, and particularly the one which asserts the state's interest in "upholding the ethical integrity of the medical profession." If a court must determine whether recognizing a particular patient's right to refuse treatment will seriously undermine the ethics of medicine, then expert testimony on that issue from a bioethicist is most certainly relevant and material.[22] An important point to note, however, is that were that question to have been entertained by a court fifty years ago, it would have been much less likely, perhaps highly improbable, that anyone would have qualified as an expert on medical ethics who did not have a medical degree, for the reasons discussed in Chapter 1. Now, with the multidisciplinary nature of bioethics well-recognized and well entrenched, bioethicists with a wide range of training and experience might be recognized by courts as possessing the requisite expertise to testify.

Yet another argument against a legitimate role for the bioethicist as an expert witness arises out of a deeper and more profound skepticism about the nature of moral knowledge, and partakes of the flavor of the Posnerian critique that we considered early in this chapter. The critique has several strains. One contention is that moral judgments lack any true objectivity, and possession of objective knowledge beyond that typically found in the lay juror is the condition precedent to being qualified to testify as an expert.[23] Another strain of the critique draws upon the emotivist theory of ethics brought to prominence in moral philosophy by Charles Stevenson.[24] Stevenson denied that there could be any logical relationship between evaluative and factual utterances. On this view, the role and purpose of evaluative utterances is not to convey true or false information, but to express personal attitudes about a particular state of affairs. Since there are no objective truths about how values and moral judgments are to be weighed and applied in specific situations, the opinions of bioethicists are in no way superior to those of the judge or jurors.[25] Hence, proffered expert testimony from a bioethicist fails the first and most essential test, that it constitute "specialized knowledge [that] will assist the trier of fact to understand the evidence or to determine a fact in issue."[26]

There is a sense in which these more fundamental arguments against expert testimony by bioethicists, in moving beyond the question of whether they *should* testify to whether they *can* (given the legal criteria for expert testimony) confuse the concept of truth with the concept of understanding. After all, there is nothing unusual about certain types of litigation, medical malpractice being a classic case in point, ultimately being characterized as a "battle of experts." Highly qualified and copiously credentialed expert witnesses called by opposite sides in the dispute will stake out diametrically opposite positions on the ultimate issue in the case, i.e., did the defendant physician's care of the patient fall inside or outside the relevant standard of care. The role of the judge or jury as trier of fact is not to determine which witness embodies *truth* with a capital "T," but which witness' testimony is more credible and persuasive, given all of the other circumstances of the case. Put another way, the trier of fact must determine which of the witnesses has demonstrated a superior understanding of medical circumstances bearing upon the appropriateness or inappropriateness of the care and treatment provided to the patient.

Those who have previously analyzed the broader issue of whether bioethics expert testimony is ever appropriate often make a distinction between two types of testimony – descriptive and normative. Even those who are most skeptical about the bioethicist as expert witness concede that there may often be a legitimate role for descriptive bioethics testimony in litigation that clearly raises bioethical issues.[27] When the expert testimony of the bioethicist is limited to what they know about ethical theory and analysis by virtue of extensive training, study, and reflection, they may serve a salutary pedagogical function in the courtroom. But, so this critique proceeds, when the bioethicist attempts to cross this strict line of demarcation into normative pronouncements about how this ethics expertise bears upon the issues in the case at bar, they have shifted from the domain of what they *know* to the domain of what they *think* or what they *believe*. To do so is to shift from a legitimate to an illegitimate role.

There is a response to this critique, and it is one with which I am in general agreement. It challenges the proposition that facts and values are invariably separable. Indeed, the entire field of applied ethics calls into question both the truth and the significance of this purported fact-value distinction.[28] This challenge was essential to the establishment of bioethics as a multidisciplinary approach that plays an evaluative role in assessing moral issues in clinical practice. We have noted previously the hidebound insistence of some prominent physicians that, for example, the decision whether or not to perform an operation on a patient is a purely medical question, about which ethicists should have nothing to say because it partakes of no normative dimensions.[29] Thus, an important role for bioethics is precisely to make the case that medicine is as much, if not more, a moral as a scientific enterprise. Even an adequate understanding of what will serve to "uphold the ethics of the medical profession" requires that we consider that medicine exists to benefit patients, and what constitutes a benefit to a patient is not just a logical or scientific deduction based upon objective data. Benefit is a normative concept, and highly dependent on the individual patient and his or her personal identity.

At least to some extent, this effort to compartmentalize bioethics expert testimony into a non-normative, descriptive or metaethical box challenges the very nature of bioethics as perhaps the most prominent form of applied ethics. It is well-known within academic philosophy that in many philosophy departments there is a caste system. Those moral philosophers who limit their focus to metaethical analysis and study of the history of ethics in Western philosophy are accepted as on roughly the same plane of the hierarchy as those working in logic or metaphysics.[30] However, those who indulge in applied ethics, e.g., bioethics and environmental ethics, are the black sheep of the department. What they are doing is considered by the traditionalists to be not "real philosophy," but public affairs, social policy, social engineering, or some other equally suspect, soft, and non-rigorous activity of limited and questionable academic rigor.[31] To some degree, then, to insist that bioethicists restrict their expert testimony to metaethics is to insist that they not be what they are – bioethicists who take their knowledge of moral philosophy and *apply* it to the issues, problems, and dilemmas posed by contemporary developments in medical science, technology, and practice. Furthermore, it is not as though some of the most revered figures in the history of philosophy disdained either the importance or the appropriateness of turning their considerable philosophical resources to real world problems. Some of the more notable instances include John Locke's writings on religious toleration, David Hume's philosophical consideration of suicide, John Stuart Mill's discussion of equality between the sexes, and Bertrand Russell's disquisitions on civil disobedience.[32] There is certainly no indication that these philosophers in any way acknowledged that in addressing these "real world" concerns of practical significance they had abandoned the domain of philosophy and hence what they had to say was of no greater merit or significance than the average person on the street (or in the jury box).

In general, the experience to date would appear to provide manifest evidence that both trial and appellate courts are perfectly capable of weighing the analyses and opinions of bioethicists, even and perhaps especially when they appear (either at trial as expert witnesses or on appeal as *amici*) and articulate conflicting views on how certain bioethical issues ought to inform the disposition of the case. In some instances the insights of one or more of the bioethicists appear to have a significant bearing on the outcome, or at least upon the court's rationale, but in others it is as though they were without any discernable influence. While certainly neither compelling nor conclusive, this evidence at least suggests a healthy state of affairs. No one, not even, I suspect, the bioethicists involved, would hope for either of the extremes, in which such professional input was either categorically rejected or uniformly ignored, or in which it invariably dictated the outcome of the legal proceedings. Nor is there any indication that bioethics as an identifiable discipline is being seriously undermined through its involvement in the legal process in some fashion unique to its nature and unlike the experience of any other discipline that has played a similar role as expert witness or friend of the court.

ENDNOTES AND REFERENCES

1. Richard A. Posner, *The Problematics of Moral and Legal Theory* (Cambridge, MA: The Belknap Press of Harvard University Press, 1999).
2. Oliver Wendell Holmes, "The Path of the Law," *Harvard Law Review* (1897), 10: 61-80.
3. Holmes, note 2, p. 63.
4. Holmes, note 2, p. 62.
5. Posner, note 1, p. ix.
6. This point is made by Ronald Dworkin in "Darwin's New Bulldog," *Harvard Law Review* (1998), 111: 1718-1738.
7. Roger B. Dworkin, *Limits – The Role of Law in Bioethical Decision Making* (Bloomington, IN: Indiana University Press, 1996).
8. Examples include Eric J. Cassell, "Moral Thought in Clinical Practice: Applying the Abstract to the Usual," in H. Tristram Engelhardt and Daniel Callahan, eds., *Science, Ethics and Medicine* (New York: Hastings on Hudson, 1976); Edmund Pellegrino and David Thomasma, *The Virtues in Medical Practice* (New York: Oxford University Press, 1993).
9. Robert M. Veatch, *The Patient -Physician Relations: The Patient As Partner* (Bloomington, IN: Indiana University Press, 1991).

10. Interestingly, Veatch's decision to transpose the traditional order of the term, from physician-patient relation to patient-physician relation, further suggests the extent of the transformation. By virtue of the doctrine of informed consent, the patient is not merely an equal partner in the relationship, but the one with ultimate authority to determine whether a proposed medical intervention will or will not take place.
11. George J. Annas, "The Dominance of American Law (and Market Values) over American Bioethics," in M. A. Grodin, ed., *Meta Medical Ethics: The Philosophical Foundations of Bioethics* (Dordrecht: Kluwer Academic Publishers, 1995), p. 83.
12. Alexander Morgan Capron, "Law and Bioethics," in Warren T. Reich, ed., *Encyclopedia of Bioethics*, (N.Y.: Simon & Schuster MacMillan, 1995), pp. 1329-1334.
13. 492 U.S. 490 (1989).
14. Annas, note 11, p. 84.
15. Ronald Dworkin, "Assisted Suicide: The Philosophers' Brief," *The New York Review of Books*, March 27, 1997, p. 41.
16. George J. Annas, Leonard H. Glantz, Wendy K. Mariner, Brief for Bioethics Professors, November 12, 1996.
17. "Bioethicists' Statement on the U.S. Supreme Court's *Cruzan* Decision," *New England Journal of Medicine* (1990), 323: 686.
18. "Dying Well In The Hospital: The Lessons of SUPPORT," Special Supplement, *Hastings Center Report* (1995), 25: S1-S36.
19. John C. Fletcher, "Bioethics in a Legal Forum: Confessions of an 'Expert' Witness," *Journal of Medicine and Philosophy* (1997), 22: 297-324.
20. Fletcher, note 19, p. 316.
21. Fletcher, note 19, p. 318.
22. This point is also made by Virginia A. Sharpe and Edmund D. Pellegrino in "Medical Ethics in the Courtroom: A Reappraisal," *Journal of Medicine and Philosophy* (1997), pp. 373-379.
23. Richard Delgado and P. G. McAllen, "The Moralist as an Expert Witness," *Boston University Law Review* (1982), pp. 869-926.
24. Charles L. Stevenson, *Ethics and Language* (New Haven: Yale University Press, 1944).
25. Giles R. Scofield, "Ethics Consultation: The Least Dangerous Profession?," *Cambridge Quarterly of Healthcare Ethics* (1993), pp. 417-426.
26. Federal Rules of Evidence §702 (1975).
27. *See, e.g.*, Giles Scofield, "Commentary: The Wizard of Oughts," *The Journal of Law, Medicine & Ethics* (2000), 28: 232-235.
28. An excellent articulation of this view in the context of an analysis of expert testimony by bioethicists can be found in George J. Agich and Bethany J. Spielman, "Ethics Expert Testimony: Against the Skeptics," *Journal of Medicine and Philosophy* (1997), 22: 381-403.
29. Sherwin Nuland, *How We Die*, (New York: Alfred A. Knopf, 1994), pp. 250-254.
30. I use the term "roughly" advisedly, for there are certainly philosophy departments in which logicians and metaphysicians would, if they had their druthers, entirely disassociate themselves from moral philosophers, even those who rigidly restrict their work to metaethics in the finest traditions of academic philosophy. In some departments this "great divide" has been made physically manifest. For example, when I was a student at the University of St. Andrews in Scotland, the philosophy department occupied a splendid stone edifice that commanded a magnificent view of the North Sea. Configured as a duplex, one side of the building housed logic and metaphysics, the other side moral philosophy. There was no way to move from one side of the building to the other without first exiting the building.
31. It is interesting to note that bioethics has been subjected to diametrically opposite critiques, as the following passage indicates:

    Among the most prominent criticisms of bioethics are that it is not serious, well-grounded scholarship; it has no well-defined and clear methodology; it lacks any solid

> conceptual foundation but is based instead on the shifting sands of moral sentiment; it is too abstractly removed from realities of clinical practice to merit being taken seriously; it is unteachable; it pursues unanswerable questions; its utility has not been demonstrated; it makes practical matters worse by confusing health care providers, policy-makers, and researchers; and it is itself ethically problematic, because it either implicitly endorses traditional values that ought to be challenged or undermines traditional values that ought to be advocated and reinforced.

Samuel Gorovitz, "Bioethics," in Lawrence C. Becker and Charlotte B. Becker, eds., *Encyclopedia of Ethics* (New York: Garland Publishing, Inc., 1992).

32. For additional examples *see* Hugo Adam Bedau, "Applied Ethics," in Lawrence C. Becker and Charlotte B. Becker, eds., *Encyclopedia of Ethics* (New York: Garland Publishing, Inc., 1992).

Chapter 10

LESSONS LEARNED AND PROSPECTS FOR THE FUTURE OF MEDICAL JURISPRUDENCE AND BIOETHICS

The general thesis of this book, stated at the beginning, is that law has had an important and generally, though not unqualifiedly, positive impact on medical ethics and medical practice. The initial sections of this chapter will review and reaffirm some of the more significant examples of law's salutary influence.

Next, we will further evaluate the extent to which at least several of the approaches to ethical analysis discussed in Chapter 2 did or perhaps should have played an influential role in the disposition of important cases in medical jurisprudence.

Finally, we will consider whether the significant role of law in the ethics of medicine and medical practice and research poses problems that should be addressed as a matter of social policy and public affairs. We will do so in the context of further considering why law has played the major role that it has, and whether it would be possible for more of that role to be taken on by bioethics and/or the organizations and entities primarily responsible for setting and maintaining standards of professionalism in medicine.

1. LAW AND THE ETHICS OF CLINICAL RESEARCH

The full story of the development and promulgation of the Nuremberg Code cannot be told with an entirely legal cast of characters. In particular, three distinguished physicians – Leo Alexander, Werner Leibbrand, and Andrew Ivy – played significant advisory roles in shaping the finished product.[1] In the course of making their respective contributions, which included testimony in direct and cross examination at the doctors' trial at Nuremberg, these individuals had to acknowledge that the ethical principles to which they sought to hold all medical researchers accountable through the Code had been violated not only by the Nazi physician defendants in those legal proceedings, but by physicians of many nationalities, including the United States.

The appalling depravity of the experiments conducted by the Nazi physicians, combined with the legal and moral authority with which the Nuremberg tribunal was endowed, may well have inspired and emboldened these medical advisors to propose a set of ethical principles for medical research that they would not otherwise have had either the inclination or the opportunity to even consider. Most certainly, they were aware that by implication the Nuremberg Code would condemn not only the activities of the Nazi physicians but some of their own professional colleagues as well. Moreover, the historical record indicates that the Nuremberg tribunal took their suggestions and actually made them more stringent in arriving at the final version.

We considered in Chapters 3 and 4 that even quite distinguished medical researchers in the United States, in the decades immediately following the promulgation of the Nuremberg Code, did not conduct their investigations in compliance with it. Thus, in a 50th anniversary reflection on the Code, one author notes that it "has not been officially adopted in its entirety as law by any nation or as ethics by any major medical association."[2] The same author credits the Code, as well as the Declaration of Helsinki promulgated by the World Health Organization in 1964,[3] for the Code's insistence that medical researchers alone cannot set the rules for the ethical conduct of research.[4]

The need for an interdisciplinary approach to the development of standards of conduct for research on human subjects, and the force of law behind such standards once issued, was reaffirmed in the United States in 1976 by the issuance of the Public Health Service-National Institutes of Health guidelines for all federally funded research.[5] Yet even today major concerns persist that the local institutional review boards (IRB's) mandated by the federal guidelines do not sufficiently scrutinize research protocols that come before them.[6]

2. THE FIDUCIARY NATURE OF THE PHYSICIAN-PATIENT RELATIONSHIP

The promulgation of the Nuremberg Code did not have nearly the impact on the treatment of human subjects in medical research as might have been anticipated. It had even less impact on the treatment of patients in the clinical setting. One might have supposed, at the very least, that the Code's first principle, informed consent, would have been considered equally relevant to, if not binding upon, the physician-patient relationship. Several reasons can be suggested for why this was not the case. First, the entire Nuremberg proceedings, and all that issued from them, were viewed as a unique and extraordinary response to the atrocities perpetrated by the Nazis. The basic, although flawed, presupposition was that physicians generally, whether their primary interest was medical research or clinical practice, were guided by an internal moral compass (or perhaps gyroscope) that acted as a fail safe device to prevent them from acting in a manner that compromised the interests of their research subjects or their patients. It is difficult to exaggerate the extent to which the beneficence imputed to the entire medical enterprise has insulated it from external scrutiny and critical review.

Medical malpractice litigation, although relatively infrequent until the 1970's, offers one of the few circumstances in which non-physicians (judges or jurors) are called upon to second-guess the appropriateness of physician conduct. But even then, with rare exceptions, they must be guided by the opinions of medical experts. What most physicians did in similar circumstances was the benchmark. The practice of most physicians in the middle of the Twentieth Century was to disclose very little to their patients about their diagnosis, prognosis, or treatment. Following the promulgation of the Nuremberg Code, there was no groundswell of support within the medical profession to reconsider the principle of nondisclosure that characterized what Jay Katz aptly referred to as *The Silent World of Doctor and Patient.* But the law, as we explored at length in Chapter 4, has always attached tremendous moral significance to the fiduciary nature of all professional relationships. Regardless of the nature of the problem prompting a person to engage a professional, the law does not presuppose that individual autonomy must be sacrificed in order to secure professional services. Moreover, because no individual or group can elevate themselves above the law, courts have been unwilling to consider themselves bound by doctrines, principles or practices another group insists should obtain in any particular state of affairs.

So it came to pass, a decade after Nuremberg, that a court invoked the Code's first principle and held that it was binding upon physicians in their interactions with patients. There is an interesting parallel between the fashioning of the Code and the opinion in *Salgo v. Stanford University.*[7] Just as several distinguished physicians served as advisors to the Nuremberg tribunal, the American College of Surgeons (ACS) filed an *amicus* brief with the court of appeals in *Salgo.* In that brief, the College actually supported the patient-plaintiff's contention that physicians had a duty to obtain not merely the patient's consent, but the patient's *informed* consent.[8]

While still not a majority view, there is a recent trend toward what might be called "judicial standard-setting." In such cases, courts in essence reject the proposition that the medical profession must be the final arbiter of what constitutes acceptable medical practice. They do so on grounds similar to the following:

> . . . customary conduct is not dispositive of and cannot overcome the requirement that physicians exercise ordinary care. . . . An emphasis on reasonable rather than customary practices . . . insures that custom will not shelter physicians who fail to adopt advances in their respective fields and who consequently fail to conform to the standard of care which both the profession and its patients have a right to expect.[9]

Such language fits comfortably within the stance taken by the physicians who worked with the Nuremberg tribunal and the ACS *amicus* brief in *Salgo*. In both instances they were calling for a standard of conduct, in medical research and medical practice respectively, that they well knew could not fairly be characterized as the "custom and practice."

It would, of course, be rank speculation to hypothesize how the doctrine of informed consent would have fared had the ACS sought to impose it as an ethical principle binding on all surgeons and, perhaps in conjunction with other national physician organizations, to impose it on all physicians. However, such speculation is directly relevant to one of the critical questions under consideration in this chapter, which is whether law or medical ethics (bioethics) should have the primary role in creating and enforcing standards of responsible professional conduct. We will take up the issue in a subsequent section, particularly the question of whether, outside of the legal domain, professional standards, particularly those that relate directly to the ethics of patient care, can be adequately enforced.

The relatively rapid and consistent adoption of the doctrine of informed consent by courts throughout the United States in the two decades following *Salgo* brought about the shift in the paradigm of the physician-patient relationship from one of paternalism to one of shared-decisionmaking. As a result of this quite revolutionary transformation, for the first time physicians began to be held to the same duty as all other fiduciaries – full and honest disclosure of all information relevant to the decisions that would be made in the course of that relationship. Medicine finally ceased to be the sole exception to the rule that fiduciaries not only owe those whose interests they serve competence, confidentiality, and fidelity, but truthfulness as well. For only medicine had adamantly insisted that patients had no entitlement to or reasonable expectation of any more information about their medical situation than their physician, in the unfettered exercise of his or her discretion, chose to provide.

Ultimately, medical ethics, or bioethics, as it had then begun to be called, followed along. Perhaps the first notable instance in which informed consent was recognized as a principle of professional ethics as well as a principle of law was the report of the President's Commission for the Study of Ethical Problems in Medicine and Biomedical and Behavioral Research. Indeed, the Commission made informed consent the centerpiece of its treatment of the broad range of issues with which it dealt: "Ethically valid consent is a process of shared decisionmaking based upon mutual respect and participation . . . Patients should have access to the information they need to help them understand their conditions and make treatment decisions . . ."[10] Thereafter, the AMA Code of Medical Ethics began to include informed consent, but as noted previously, with the curious reference to it as matter of social policy rather that a core ethical responsibility of the physician.

Old and well-entrenched beliefs die hard, even within the learned professions. One such belief, common among most earlier generations of physicians and still far from completely eradicated, is that medical decisions such as whether a surgical procedure should or should not be performed, are clinical, objective and non-normative in nature. As such, they should be made by qualified professionals only, i.e., physicians. Those who argue otherwise – patient advocates, lawyers and judges, bioethicists – are seen by those in the grip of this belief as ignorant and malicious intermeddlers in medical matters.[11] Worse still, patients who doggedly refused to consent to treatment that was deemed by their physician to be indicated, and perhaps even necessary, had their decisional capacity challenged. Failure or refusal to agree with one's physician was prima facia evidence of a mental problem. On the other hand, patients who might demonstrate any manner of bizarre ideation, yet go along with the recommendations for treatment, might never have their decisional capacity called into question.

3. THE PATIENT'S RIGHT TO DECIDE

Many of the other key transformations in the ethics of healthcare decisionmaking followed from the patient's legal right to informed consent. Information, properly presented, yields knowledge, and the recognized right to utilize that knowledge to make decisions and to act upon them is power. Recognition of the right to give an informed consent, and its reciprocal, the right to give an informed refusal, created a balance of power in the physician-patient relationship that had never before existed. The right to decide – to give or refuse consent following a disclosure of the relevant information – seems implicit in the doctrine of informed consent. Consequently, it is remarkable how much litigation in the decades of the 1970's and 1980's was necessary before that right was widely and routinely recognized to extend to life-saving or life-extending medical interventions, and that this right could be conveyed from one person to another (as surrogate) through a written instrument (advance directive).

During this period, both medical jurisprudence and bioethics struggled with the moral implications of decisions about treatment that were likely to result in the patient's death. In the documentary film "Dax's Case," a retrospective look at Dax Cowart's unsuccessful efforts to refuse the many months of excruciating treatment for the third degree burns that he sustained over two-thirds of his body, his physicians insist that what Cowart was demanding was that they kill him. Similar sentiments were expressed by those physicians who subjected Elizabeth Bouvia, more than ten years later, to artificial feeding over her express refusal. Neither Cowart nor Bouvia ever were, nor could they have been, found to lack decisional capacity. But they were both viewed by those inflicting treatment upon them as trying to commit suicide and to secure the assistance of their physicians in doing so. Thus, the "right to decide" only existed, from this perspective, in cases where the patient's life was not in serious jeopardy. In "Dax's Case," the patient's mother, a devoutly religious person to whom his physicians looked for consent for his treatment, observes that if God had taken her son she could have come to terms with that. But she could not accept his own choice of death. Her remarks reflect a common perception among lay persons and health professionals alike with regard to the care of critically ill patients: to provide any and all treatment available, even in the face of the patient's refusal or a very poor prognosis is just "playing doctor;" but to respect the patient's refusal of treatment or to withhold or withdraw treatment because of the likelihood that it will not succeed, is "playing God."

What law and bioethics did, working now in a more coordinated, or at least consistent, fashion, was to draw important distinctions between killing and letting die. Patients with serious, irreversible, and ultimately fatal medical conditions can, through the rejection of life-sustaining medical interventions, exercise a measure of control over the time and manner of their death. This is true not only with regard to cancer patients who refuse surgery and/or chemotherapy, but also for patients with congestive heart failure who refuse drug therapy or chronic obstructive pulmonary disease who refuse mechanical ventilation. In the 1970's and 1980's a joint consensus of state and federal courts and the bioethics community maintained that a patient's refusal to undergo these medical procedures did not involve the medical profession in "killing" these patients in any active or morally problematic sense. Rather, through the exercise of and respect for patient autonomy, the natural progression of the disease process is allowed to go forward unhindered by medical measures. While, as a matter of personal or professional psychology, it may feel as though the physician is killing the patient by failing to provide medical interventions that will forestall death, the prior recognition by the courts of the patient's right to informed consent presupposes a right to informed refusal as the logical corollary.

The complicating factor, as we discussed in Chapter 8, has been the invocation of the four state interests that were countervailing to the patient's right to decide whether to accept all, some, or none of the medically indicated treatment. Because these state interests were generally asserted in actual legal cases – sometimes directly by the state and sometimes by healthcare professionals as proxy – the law has had a much more significant role than bioethics in determining the outcome of these disputes.[12] As we noted in Chapter 9, however, since one of the four interests is said to be upholding the ethical integrity of the medical profession, there was a genuine opportunity for bioethicists to contribute to the debate as to whether allowing a patient to die consistent with his or her wish to decline further treatment could be considered, in any rational and principled manner, to compromise the ethics of the profession. Of considerable influence in this regard was the report of the President's Commission, which stated:

> The voluntary choice of a competent and informed patient should determine whether or not life sustaining therapy will be undertaken, just as such choices provide the basis for other decisions about medical treatment . . . Health care professionals serve patients best by maintaining a presumption in favor of sustaining life, while recognizing that competent patients are entitled to choose to forgo any treatments, including those that sustain life.[13]

As recently as 1990, the question of whether artificial nutrition and hydration constituted a medical procedure that could be refused was still being debated. Hopefully, the Supreme Court's statement in the *Cruzan* decision that the invasive nature of the means by which it is supplied makes it a medical procedure that can be declined has finally put this issue to rest.

4. PROSPECTIVE AUTONOMY AND ADVANCE DIRECTIVES

Both law and bioethics have a checkered history when it comes to the prospective refusal of life-sustaining treatment through the use of advance directives. While they are both supportive of the use of advance directives, they also tolerate a tremendous amount of passivity and indifference on the part of the medical profession. Physicians are under no legal or ethical obligation to initiate or even support the discussion and execution of advance directives with their patients. Furthermore, to date physicians have been subject to neither disciplinary action nor legal liability for their failure to respect advance directives when they are in place. In the tidal wave of commentary that followed the publication of the SUPPORT data, only one person had the temerity to suggest that a physician who refuses to respect a patient's directive or to disengage from the physician-patient relationship should be subject to legal sanctions.[14] A few prominent bioethicists, as we noted in Chapter 5, have even advanced the position that advance directives lack moral authority because the patient who requires life-sustaining treatment is a different person than the competent individual who executed the directive.[15] The medical maxim to "err on the side of life" continues to operate with undiminished vitality when advance directives or patient proxies equivocate in any fashion. Conservative courts have persistently imposed unrealistically high standards of specificity on written, but particularly upon oral, directives purporting to refuse life-sustaining treatment.[16] The Supreme Court decision in *Cruzan* essentially provided the states with *carte blanche* to be as restrictive as they care to be. Moreover, the Court has shown a distinct disinclination to revisit, and perhaps clarify, issues that remain after *Cruzan*.[17]

What we have at the present time is aggressive, invasive, life-sustaining interventions as the default position when the patient lacks decisional capacity and advance directives are either ambiguous or nonexistent. It is not clear whether this default position is tied to the purported interest of the state in preserving life regardless of its quality, or whether it is supposed to reflect a social consensus. The former appears more plausible, since the impetus for advance directive legislation was a strong public consensus that most people would not wish to have their lives prolonged in a profoundly and irreversibly debilitated condition. Except for instances in which the evidence as to a particular patient is equivocal or contradictory, patients who have permanently lost the capacity to interact with their environment (advanced dementia, minimally conscious and persistent vegetative states) or who lack decisional capacity and have a terminal or end-stage illness and who have previously stated a disinclination to have their life prolonged, should not have treatment foisted upon them. The burden of proof should be on those seeking to err on the side of life, and they should not be able to meet that burden without clear and convincing evidence that the patient in question really wished treatment under such circumstances.

5. PHYSICIAN-ASSISTED SUICIDE, EUTHANASIA, AND AGGRESSIVE PALLIATIVE MEASURES

The debate over physician-assisted suicide (PAS) and euthanasia came to a head in the United States in the decade of the 1990's. It would not be an exaggeration to state that Americans are at least as morally divided over this issue as they are on abortion. Two very influential federal circuit courts of appeal found a constitutional right to PAS on the part of competent, terminally ill patients in different clauses of the Fourteenth Amendment. While the Supreme Court saw it differently, it also readily acknowledged that there was nothing in the Constitution that would preclude a state like Oregon from legalizing and regulating such a practice. While jury after jury refused to convict Dr. Jack Kevorkian of a crime when he provided his assistance to patients who wished to hasten their deaths, he was promptly convicted and imprisoned when he crossed the line from PAS to voluntary active euthanasia by administering a lethal injection to an ALS patient who could not self-administer the death inducing agent.

There are now 3 full years of data on the Oregon Death with Dignity Act.[18] While that still may be an insufficient basis for drawing a firm conclusion that the purported risks of a slippery slope have been avoided, it does at least suggest that neither the citizens of Oregon nor their physicians are now in the grip of a culture of death. For example, during the first 3 years 96 prescriptions for lethal doses of medication were written. Of those patients, 69 actually utilized the prescription to end their lives. Roughly 75% of those obtaining a lethal prescription were enrolled in a hospice program, and most had some form of health insurance.[19] As this book goes to press, no other state has elected to legalize and regulate physician-assisted suicide. Moreover, while the Pain Relief Promotion Act (intended to make the writing of a lethal prescription a violation of the federal Controlled Substances Act) has not been reintroduced in the current Congress, it remains to be seen whether the new administration will seek to achieve the same result through an administrative ruling by the Department of Justice. If writing a lethal prescription were to become a federal offense, the impact on PAS in the United States would be profound. It would also, of course, run directly counter to all of the reverential rhetoric of the Supreme Court in the PAS decisions about the salutary features of "the laboratory of the states."

Perhaps the most important aspect of the so-called "right to die" litigation of the last 30 years is the creation of meaningful distinctions between killing patients and appropriately participating in their dying. While such distinctions remain contentious on certain levels of philosophical discourse, at least some activities now appear to be beyond reasonable doubt. For example, in the previously mentioned videotape of 30 years ago one of Dax Cowart's physicians adamantly insisted that by requesting the discontinuation of all therapeutic interventions Cowart was asking to be killed. Today, the law is clear and unequivocal that withdrawing treatment necessary to sustain life pursuant to a patient's or a duly designated surrogate's wishes is neither homicide nor assisted suicide. Regrettably, however, this does not mean that health care professionals no longer labor under the misapprehension that it is.[20] The continuing paucity and brevity of courses on medical jurisprudence in medical schools insures that myth and misinformation continue to be passed along from one generation of physicians to another through the informal (hidden) curriculum.

6. ETHICAL THEORY AND MEDICAL JURISPRUDENCE

It should not surprise us that the ethical principle that is most consistently cited in medical jurisprudence is respect for patient autonomy, inasmuch as it is so closely aligned with core legal and political principles of individual liberty and privacy. The courts have been much less receptive to some of the more esoteric approaches to bioethical issues considered in Chapter 2, such as virtue theory, narrative ethics, feminist or care ethics. Persuasive critiques can be offered of many of the key Supreme Court decisions – *Roe, Casey, Cruzan, Glucksberg* – that the voice of the patient and the family were virtually silenced by the justices' attention to the political claims of conflicting interest groups, especially those of the health professions and the right-to-life advocates.[21] Justice Blackmun, writing for the Court in *Roe*, was incredibly solicitous of the medical profession, exceedingly thorough in his explication of abortion in history, and determined to resolve the issue of the ontological status of the fetus vis-a-vis the Constitution. Neither he nor his colleagues on the Court expressed any particular interest in or concern for the plight of pregnant women who could not access abortion services. *Casey* is perhaps the most philosophical of the major Supreme Court decisions with bioethical implications, touching, however briefly, on what it means to be a person in a free and democratic society. One might even suggest that the voices of pregnant women were actually heard when the Court struck down the Pennsylvania law requiring spousal notification of the intent to obtain an abortion. While a feminist or care ethic might have strongly militated toward a different outcome than the Court reached on a number of the other provisions of Pennsylvania abortion law, it would be unreasonable to suggest that any ethical stance must trump all other considerations in every case.

In *Cruzan,* the Opinion of the Court, unlike those of the dissenters, was much more concerned about other hypothetical patients – those without loving families – and with the interests of the state, than it was in the plight of Nancy Cruzan and her family. As with a number of state supreme court decisions, there is a profoundly anti-narrative tone in the Court's language. The clear and convincing evidentiary standard that was upheld by the Court and imposed in jurisdictions such as Missouri, Michigan, and New York, is both overly legalistic and disturbingly unrealistic, as many dissenting opinions in the cases have heatedly pointed out. Most people do not communicate about the most intimate and personal details of their living and dying in documents that meet rigid technical requirements for prescience and specificity; but they do communicate. By imposing unreasonably high expectations on the manner and mode of communication that will be given credence, courts effectively turn a deaf ear to the voices of patients and to those in the closest caring relationships with them.

In the PAS cases, the voices of the named patients and their physicians were silenced by the disingenuous ploy of treating the challenges to the statutes of New York and Washington as facial, rather than what they were, which was "as applied" to those particular patients. By treating the challenges in this way, the Court could completely disregard the narrative and the plight of these plaintiffs. Indeed, as noted earlier, Justice Stevens went so far as to make the truly bizarre suggestion that he might vote differently if he were presented with a case in which the patient's situation was more like that of the actual plaintiffs in these cases. Theatre of the absurd presents no more incredible scenes than this.[22]

One can certainly argue that the dissenting opinions in *Cruzan*, and to some extent the concurrences in *Glucksberg* and *Quill*, were an effort on the part of those justices to reassure the citizens of the United States that they did hear the tragic personal narratives of the plaintiffs, and that they did care about the personal consequences of the Court's ruling. The five justices who joined in concurring opinions in the PAS cases were clearly concerned about the plight of dying patients experiencing significant pain and suffering. They sought to reassure Americans that the Court would not countenance the erection or maintenance of legal barriers to effective pain relief. However, while one can criticize these justices for their incredible naivete about the actual barriers to pain relief, a more serious critique would focus on their failure to recognize the needs of those with intractable suffering who seek to determine for themselves, in concert with a compassionate physician, what constitutes a dignified death.

A genuine embrace by the Court of even the most rudimentary elements of narrative, feminist and care ethics might have dramatically changed the focus and tone of the opinions, if not the outcome of the cases. Without an exhaustive review of the *amicus* briefs filed in these cases, we cannot state with certainty whether or not narrative, feminist and care ethical analyses were presented to the Court, but we do know that if those seeds were sown, they did not find fertile ground in many of the justices. In fairness, we must also guard against any presupposition that these particular ethical perspectives would invariably support judicial recognition of a broad right to abortion, withdrawal of life-support by surrogates, or PAS. One could certainly construct an argument against abortion rights, surrogate decisionmaking, and PAS that partakes of strands of narrative, feminist, or care ethics.

Bioethicists can and should be much more zealous in their advocacy of these marginalized ethical perspectives when they have a salutary role to play in humanizing and personalizing judicial decisionmaking. Indeed, that may be one of the more important roles for bioethics in the Twenty-First Century. But as we have seen from our review of some *amicus* briefs in Chapter 9, in order to do so the briefs will need to look a lot less like traditional legal briefs and a lot more like carefully argued and copiously documented position statements. The burden of persuasion will be that particular ethical theories or perspectives are essential to a proper analysis of the legal issues and the implications and ramifications of finding in favor of one party to the litigation rather than another. Appellate courts have in the past shown themselves to be open to briefs that provide the judges with information from the hard sciences and even the social sciences when it has clear implications (materiality and relevance) for a proper resolution of the issues presented by the case.

7. CAN WE AVOID THE LEGAL ENFORCEMENT OF MEDICAL ETHICS?

At first glance, such a question sounds like a throwback to an earlier century, when medical matters were virtually the exclusive province of physicians. Whether or not a patient had been treated properly by a physician in any regard whatsoever was within the exclusive province of the profession of medicine to determine. Even in the most basic medical malpractice claim, conformity with the custom and practice of physicians in the locality virtually insulated the defendant physician from liability. In addition to its well-known antipathy to lawyers based upon their role in filing medical malpractice suits on behalf of injured patients, the medical profession, as we have noted in previous chapters, strenuously resisted the insinuation of the law or governmental entities into any aspect of medical research and clinical practice. The only exceptions were in those instances, most particularly noted in Chapters 7 and 8, in which physicians sought to insulate themselves from liability through the imprimatur of the legislatures or the courts regarding abortion policy and end-of-life care.

The medical profession has assiduously cultivated the perception that the principle of beneficence reigns supreme, and consequently there is rarely, if ever, an instance in which the individual physician is not pursuing the patient's best interests. But as we have explored at great length, particularly in Chapters 4 and 8, a physician's view of the patient's best interests is an extremely narrow, rigid, and medicalized view, leaving little room for the thoughtful consideration of the patient's personal values, goals, hopes, and fears. Moreover, while attendance at any departmental morbidity and mortality conference readily confirms the capacity of physicians to be highly critical of one another within the hierarchical structure of the academic medical center on matters of strictly technical performance, there is a diametrically opposite resistance to critique one another, individually or collectively, on matters of ethics. This is particularly the case when there is any possibility that such a critique might reach beyond the confidential precincts of a particular clinical setting.

The American Medical Association Code of Medical Ethics, to which we have referred at various points throughout this book, is impressive. Yet, it is at best a hortatory document. As we noted in Chapter 8, the SUPPORT data strongly suggests that the conduct of many physicians violated (by act or omission) one or more provisions of the Code. Nevertheless, none of those physicians has suffered any consequences as a result. Similarly, existing codes of medical ethics neither prospectively deterred nor retrospectively sanctioned those medical researchers who flagrantly violated the first principle of the Nuremberg Code in research at Tuskeegee, Cincinnati, and other locations throughout the United States.[23] This suggests that, as physician Leon Kass observed, paraphrasing Aristotle in *The Nichomachean Ethics*, "ethics is impotent without politics – without careful attention to law and custom and the ordering of civic life."[24]

One topic in bioethics, a detailed analysis of which is beyond the scope of this book, is the role of institutional ethics committees. It is difficult to generalize about such committees because they do vary widely from one institution to another. However, most have consciously sought to avoid being characterized as the moral police. They may welcome opportunities to participate in the review and discussion of ethically complex clinical cases, but they almost without exception decline to decide ethical disputes (as opposed to offering opinions or suggestions). Nor, so far as I know, have any of them sought or received authority within the institution to impose sanctions for unethical practice. At most, such committees may recommend changes in institutional policy based on ethical concerns that have arisen. But even in such instances, disciplinary action of health care professionals for actions in violation of such policies would be within the purview of the medical staff or the institutional board of directors. Furthermore, the institutional clout of ethics committees tends to be directly proportional to the number of physicians who actively participate as members, if not leaders. The greater the influence of the medical staff on the ethics committee, the less likely that the committee will attempt to impose its own views on what constitutes ethical practice.[25]

The chronicle of the law's involvement in bioethical matters has been one of responding to problems that had been identified yet tolerated. I characterize it as one of filling the moral vacuum that the law tends to abhor.

Whether the issue is informed patient consent to undergo medical treatment or to participate in clinical research, recognition of the concomitant right to refuse treatment or to secure adherence to an advance directive, or to receive an appropriate screening exam and stabilizing treatment of an emergency medical condition before transfer to another institution, the law interceded when health care institutions and professionals failed to effectively address the problems (or in some cases denied the existence of a problem). The extent to which the law's intervention actually produced meaningful reform was usually directly proportional to the sanctions imposed for failure to act in conformity with its provisions.

George Annas has suggested: "One may have to be subversive to be ethical in American health care institutions."[26] Of course, the very meaning of subversion is to overthrow prevailing institutions or practices. The idea of an institutional committee having, within its charge, subverted certain practices that have become commonplace within the institution, appears patently absurd. The more effective the committee becomes in such a role, the more imperiled it will be. That is why the law, as the perennial outsider to the customs and practices of health care institutions and professionals, is perfectly situated to support the individuals who take on the subversive role and seek to reform policies and practices that violate existing or emerging ethical standards. Bioethics has many opportunities, as we noted in Chapter 9, to be an active and influential player in the various legal arenas. But ultimately, when the controversy seems otherwise intractable, resort to the force of the law to impose sanctions and establish priorities is essential.

ENDNOTES AND REFERENCES

1. Evelyne Shuster, "Fifty Years Later: The Significance of the Nuremberg Code," *New England Journal of Medicine* (1997), 337: 1436-1440.
2. Shuster, note 1, p. 1439.
3. 21 C.F.R. § 312.120(b)(4) (1993).
4. Shuster, note 1, p. 1440.
5. Mark S. Frankel, "The Public Health Service Guidelines Governing Research in Human Subjects," *George Washington University Program of Policy Studies in Science and Technology*, Monograph No. 10 (Washington, D.C., 1972), pp. 20-21.
6. Office of the Inspector General, "Institutional Review Boards: A Time for Reform," OEI-01-97-00193, June 1998.
7. 317 P.2d 170 (Cal. App. 1957).
8. Jay Katz, *The Silent World of Doctor and Patient* (New York: Free Press, 1984), p. 60. Katz notes that the entire paragraph of the opinion in which the term " informed consent" is first used and recognized "was adopted verbatim, and without attribution, from the *amicus curiae* submitted by the American College of Surgeons . . . ."
9. Nowatske v. Osterloh, 543 N.W.2d 265 (Wis. 1996).

10. President's Commission for the Study of Ethical Problems in Medicine and Biomedical and Behavioral Research, *Summing Up: The Ethical and Legal Problems in Medicine and Biomedical and Behavioral Research* (Washington, DC: U.S. Government Printing Office, 1983), pp. 20-21.
11. Sherwin Nuland, M.D., *How We Die* (N.Y. Alfred A. Knopf, 1994), pp. 252-253.
12. Ben A. Rich, "The Assault on Privacy in Healthcare Decisionmaking," *Denver University Law Review* (1991), 68: 1-55.
13. President's Commission for the Study of Ethical Problems in Medicine and Biomedical and Behavioral Research, *Deciding to Forgo Life Sustaining Treatment* (Washington, D.C.: U.S. Government Printing Office, 1983), p. 23.
14. George J. Annas, "How We Lie," Special Supplement, *Hastings Center Report* (1995), 6: S12-S14.
15. *See, e.g.*, Rebecca Dresser, "Life, Death, and Incompetent Patients: Conceptual Infirmities and Hidden Values in the Law," *Arizona Law Review* (1986), 28: 377-405.
16. *See, e.g.*, *In re* Westchester County Med. Center, 531 N.E.2d 607 (1988).
17. This was especially true of the unsuccessful petition for *certiorari* by the petitioners in In re Martin, 538 N.W.2d (Mich. 1995).
18. The data and analysis provided in Oregon's Death with Dignity Act Annual Report 2000 can be accessed at www.ohd.hr.state.or.us/chs/pas/arresult.htm.
19. See note 17.
20. Alan Meisel, Lois Snyder, Timothy Quill, "Seven Legal Barriers to End-of-Life Care: Myths, Realities, and Grains of Truth," *Journal of the American Medical Association* (2000), 284: 2495-2501.
21. One of the more recent and extensive of these critiques is Susan M. Behuniak, *A Caring Jurisprudence: Listening to Patients at the Supreme Court*, (Lanham, MD: Rowman & Littlefield Publishers, Inc., 1999).
22. Behuniak offers a much more sympathetic account of Justice Stevens' concurring opinion in *Glucksberg* and *Quill*, but she does so having bought into the Court's position that these were genuinely facial challenges to the statutes, when a careful reading of the Ninth Circuit's opinion in *Glucksberg* clearly indicates that the plaintiffs were making an "as applied" challenge to the Washington statute. See note 39, pp. 144-145.
23. *See, e.g.*, Keith Schneider, "Scientists Are Sharing the Anguish over Nuclear Experiments on People," *New York Times*, March 2, 1994, p. A9.
24. Leon Kass, "Practicing Ethics: Where's the Action?" *Hastings Center Report* (1990), 20: 5-12, p. 10.
25. One prominent bioethicist recently wrote: "Too many of us who work in hospitals or medical centers run the risk of simply ratifying physicians' beliefs. We must preserve our independence and our outsider status." Gregory E. Pence, "Setting A Common, Careful Policy for Bioethics," *Chronicle of Higher Education* (2001), 47(18): B20.
26. George J. Annas, "The Dominance of American Law (and Market Values) Over American Bioethics," in *Meta Medical Ethics: The Philosophical Foundations of Bioethics*, Michael A. Grodin, ed., (Dordrecht: Kluwer Academic Publishers, 1995).

INDEX